The Biological Foundations of Action

Philosophers have traditionally assumed that the difference between active and passive movement could be explained by the presence or absence of an intention in the mind of the agent. This assumption has led to the neglect of many interesting active behaviors that do not depend on intentions, including the "mindless" actions of humans and the activities of non-human animals. In this book Jones offers a broad account of agency that unifies these cases. The book addresses a range of questions, including: When are movements properly attributed to whole agents, rather than to their parts? What does it mean for an agent to guide its action? What distinguishes agents from other complex systems? What is the relationship between action and adaptive behavior? And why might the study of living systems be the key to understanding agency?

This book makes an important contribution to current philosophical debate on the nature and origins of agency. It defines action as a uniquely biological process and recasts human intentional action as a specialized case of a broader and more common phenomenon than has been previously assumed. Uniting findings from philosophy, cognitive science, psychology, biology, computer science, complexity theory and ethology, this book will be of interest to students and scholars working in these areas.

Derek M. Jones is Assistant Professor of Philosophy and Director of Cognitive Science at the University of Evansville, Evansville, Indiana, USA.

History and Philosophy of Biology
Series editor: Rasmus Grønfeldt Winther | rgw@ucsc.edu |
www.rgwinther.com

This series explores significant developments in the life sciences from historical and philosophical perspectives. Historical episodes include Aristotelian biology; Greek and Islamic biology and medicine; Renaissance biology; natural history; Darwinian evolution; nineteenth-century physiology and cell theory; twentieth-century genetics, ecology, and systematics; and the biological theories and practices of non-Western perspectives. Philosophical topics include individuality, reductionism and holism, fitness, levels of selection, mechanism and teleology, and the nature-nurture debates, as well as explanation, confirmation, inference, experiment, scientific practice, and models and theories vis-à-vis the biological sciences.

Authors are also invited to inquire into the "and" of this series. How has, does, and will the history of biology impact philosophical understandings of life? How can philosophy help us analyze the historical contingency of, and structural constraints on, scientific knowledge about biological processes and systems? In probing the interweaving of history and philosophy of biology, scholarly investigation could usefully turn to values, power, and potential future uses and abuses of biological knowledge.

The scientific scope of the series includes evolutionary theory, environmental sciences, genomics, molecular biology, systems biology, biotechnology, biomedicine, race and ethnicity, and sex and gender. These areas of the biological sciences are not silos, and tracking their impact on other sciences such as psychology, economics, and sociology, and the behavioral and human sciences more generally, is also within the purview of this series.

Rasmus Grønfeldt Winther is Associate Professor of Philosophy at the University of California, Santa Cruz (UCSC) and Visiting Scholar of Philosophy at Stanford University (2015–2016). He works in the philosophy of science and the philosophy of biology and has strong interests in metaphysics, epistemology, and political philosophy, in addition to cartography and geographic information systems (GIS), cosmology and particle physics, psychological and cognitive science, and science in general. Recent publications include "The Structure of Scientific Theories," *The Stanford Encyclopaedia of Philosophy* and "Race and Biology," *The Routledge Companion to the Philosophy of Race*. His book with University of Chicago Press, *When Maps Become the World*, is forthcoming.

Published

Romantic Biology, 1890–1945
Maurizio Esposito

Natural Kinds and Classification in Scientific Practice
Edited by Catherine Kendig

Organisms and Personal Identity
Individuation and the Work of David Wiggins
A.M. Ferner

The Biological Foundations of Action
Derek M. Jones

Forthcoming

Darwinism and Pragmatism
William James on Evolution and Self-Transformation
Lucas McGranahan

The Biological Foundations of Action

Derek M. Jones

Routledge
Taylor & Francis Group

LONDON AND NEW YORK

First published 2017 by Routledge

2 Park Square, Milton Park, Abingdon, Oxfordshire OX14 4RN

52 Vanderbilt Avenue, New York, NY 10017

Routledge is an imprint of the Taylor & Francis Group, an informa business

First issued in paperback 2019

British Library Cataloguing in Publication Data
A catalogue record for this book is available from the British Library

Library of Congress Cataloging-in-Publication Data
Names: Jones, Derek M., 1980– author.
Title: The biological foundations of action / Derek M. Jones.
Description: Milton Park, Abingdon, Oxon ; New York, NY : Routledge,
    2017. | Includes bibliographical references.
Identifiers: LCCN 2016006781 | ISBN 9781848935341 (hardback) |
    ISBN 9781315559698 (e-book)
Subjects: LCSH: Act (Philosophy) | Agent (Philosophy)
Classification: LCC B105.A35 J66 2017 | DDC 128/.4—dc23
LC record available at https://lccn.loc.gov/2016006781

ISBN: 978-1-8489-3534-1 (hbk)
ISBN: 978-0-367-35860-0 (pbk)

Typeset in Times New Roman
by Apex CoVantage, LLC

Contents

Figures

Acknowledgments

I am indebted to a number of many talented philosophers and scientists for their support and guidance throughout this project. Frederick Schmitt, Randall Beer, Kirk Ludwig, Timothy O'Connor, Rasmus Winther and four anonymous reviewers offered comments that improved this manuscript significantly. I am grateful to my departmental colleagues at the University of Evansville – Anthony Beavers, Lisa Kretz, Dick Connolly, Dianne Oliver, Jim Ware and Valerie Stein – for their insight and moral support. I also thank my students at the University of Evansville, some of whom were an audience to earlier versions of these ideas (a captive audience, but a gracious one nonetheless).

Thanks as well to my sons and favorite agents, Colin and Graham Jones, who remain an endless source of philosophical material, and to Marisa Jones for her unwavering (and at times unjustified) confidence in me. Without her support I might not be doing very much at all.

Introduction

Does nothing lie between a corpse-like graven image and a vehicle for reason?

Brian O'Shaughnessy[1]

The task of the philosophy of agency is to understand how things like us fit into the rest of the universe. Most of the events that take place in the universe are mere *happenings*. Happenings include such events as ionization, precipitation, decomposition, and erosion. These events may be complex and significant but they are not *doings*. They constitute the background against which active systems such as ourselves – agents – carry out their activities. We attach far greater significance to the events we describe as doings or, less clumsily, as *actions*. They range from the simple (walking, sitting, raising a limb) to the complex (driving, marrying, running for office). Actions are unique events because it takes an agent to produce action, and agents seem to be exceedingly rare. Moreover, an event must be produced in the right way to count as an action. Reflection on these facts raises several related questions: What distinguishes agents from the rest of the organized matter in the universe? What gives action this privileged status in the grand causal scheme of things? Is it limited to the mindful behavior of *homo sapiens sapiens*, or might it arise elsewhere?

Philosophers have traditionally addressed these questions by examining cases of intentional action, with the planning, intending, reasoning human cast as the prototypical agent. But although planned actions are unquestionably doings and human beings are undeniably agents, we should worry that by focusing on such sophisticated targets we lose sight of broader commonalities. The intellectualist preoccupation with plans and intentions can blind us to the existence of a large class of behaviors that, although falling short of what we might call 'full-blooded' intentional action, nonetheless share the core characteristics of action. When I absentmindedly drum my fingers on my desk while reading, I am *doing* something, and this is so even if I did not plan, intend, or have any reason whatsoever to do it. The behavior does not merely happen to me – indeed, I would be horrified to discover that my fingers were drumming *themselves*. By contrast, I know that my heart is beating without my having much control over it, and I am content to let it work without my guidance.

The traditional approach also neglects the cases of animals that seem quite active despite lacking anything resembling human reasons for action. It seems unlikely that spiders token the same types of intentional states that we do, but we can nonetheless recognize a difference between when a spider moves its leg and when its leg is moved by an outside force. Indeed, the distinction between activity and passivity may hold for very basic systems – we say that the amoeba engulfs the cyst – engulfing being something the amoeba does – but that the plasmolysis taking place at its membrane is a passive process. Our intuitions suggest a response to O'Shaughnessy's question: a range of interesting cases fall between graven images and vehicles for reason, and a complete philosophy of agency will account for them.

So much the worse for our intuitions, some might argue. The rich literature on deliberative agency addresses a genuine phenomenon of interest. Why should we insist that these simpler 'actions' are anything other than ersatz cases, interesting only insofar as they approximate to the genuine article? In what follows I hope to answer this challenge by developing a plausible unifying account of action and agency.

There is historical precedent for this sort of foundational investigation. Cognitive science has benefited from a shift in focus from the most sophisticated forms of human cognition to the means by which very simple systems navigate their unpredictable environments in real time. With an understanding of those systems in hand, theorists can attempt to understand how the adaptive strategies of simple agents could be scaled up to the level of human thinkers. Unsurprisingly, this shift has generated renewed interest in the origins of life among cognitive scientists. This bottom-up approach has launched exciting new research programs in situated, embodied, and enactive cognitive science. My primary motivation for writing this book is the idea that the same methodological strategy should be applied to the study of action and agency.

I offer a general analysis of this broader form of action in Chapter 1, which will set up the problems to be investigated for the rest of the book. Agency in its broadest sense, which I call *primitive agency*, is characterized by whole-system guidance of coordinated behavior. But what does the guidance relation entail, and what could it mean for a *whole system* to guide its behavior? I survey several candidate answers to these questions in Chapters 2 and 3 and conclude that they cannot be answered from a mechanistic, reductionist perspective. We can only understand whole-system guidance from the perspective of systems theory. I survey the general features of that perspective and their application to the self-organization of behavior in Chapter 4.

The shift to a systems-theoretic perspective generates unique problems. Agents are self-organized systems that both comprise and are composed of other self-organized systems. In a world of flux, what guides us in the task of distinguishing agential systems from other patterns of organized activity? One response is that we are guided by little more than explanatory convenience; given our interests, we may interpret patterns as we see fit, but there is no fact of the matter as to which complex systems constitute agents. In Chapter 5 I reject this view in favor of the view that living systems are uniquely organized so as to distinguish *themselves* from their environments as unified wholes. These self-producing systems both constitute objective unities and serve as the primordial springs of objective value and meaning in the world. Here the idea that agency is an essentially normative form of behavior will play a significant role.

Life is necessary but insufficient for agency. An agential system must also be able to modify its behavior in accordance with the norms of survival engendered by its organization. In short, it must be *adaptive*. I discuss and refine one plausible account of adaptivity in Chapter 6. The embedded character of the sensorimotor animal comes with unique normative demands. These demands, involved with the investment of energy in the construction and maintenance of adaptive sensorimotor behavior, are the hallmark of primitive agency and constitute the foundations of action in general.

The resulting picture depicts the human agent as a specialized subkind of a broader class of biological systems. It allows us to recognize what is special about humanity while respecting the deep continuities that run through the biological world. Agency remains a rarity in the universe, but if the following line of argument holds, human agents have a bit more company than we have traditionally thought.

Note

1 B. O'Shaughnessy, *The Will: A Dual Aspect Theory* (2nd ed., Cambridge: UK: Cambridge University Press, 2008), pp. 54–55 note 2.

1   On the need for a theory of primitive action

Activity, passivity, and intellectualist motivations

To begin our discussion of action, consider the distinction between active and passive movement implicit in the introductory discussion of doings and happenings. I act when I leap from a cliff, but I do not act in being thrown from that same cliff. I am active in the former case because my behavior is settled (at least partially) by me. I am passive in the latter case because my behavior is settled entirely by external forces. But the distinction between active and passive movement cannot be captured simply by distinguishing movement generated from within the agent from movement generated from without. A bomb placed within an agent does not produce action by exploding. Nor is it enough to require the internal cause to be a properly functioning part of the agent, for peristaltic contractions in the intestines and the cardiac cycle, although perfectly natural, are not actions. Nor will a broad appeal to neuroanatomy suffice, because seizures have neural origins but nonetheless fail to qualify as actions.

One early attempt to capture the distinction between active and passive movement would be to distinguish those movements that are describable as either intentional or unintentional from those that are not. I intentionally raise my arm; I unintentionally annoy the stranger behind me by raising it. By contrast, I do not writhe intentionally or unintentionally when in the grip of a seizure. This appears to capture the active/passive distinction: Arm-raising and stranger-annoying are things that one *does*. Seizures are things that *happen to* their sufferers. Viewed this way, the problem of discriminating action from nonaction may be treated as the problem of discriminating movements that are (un)intentional from movements that are not, where the relevant difference hangs on the presence or absence of certain 'intentionalizing' states of the agent. Our philosophical attention then shifts to the matter of what those states must be.

Ludwig Wittgenstein famously posed a question regarding the difference between one's raising the arm and one's arm simply going up.[1] The difference is often thought to be traceable to a difference in causal antecedent: assuming that such a subtraction of the bodily movement from the act is coherent, the remainder must be some prior state. That prior state is usually taken to be a form of *intention* that serves as the cause of the movement. When I intend to raise my arm and my arm goes up, then, given that the right sort of causal connection is obtained between the intention and the movement, it seems right to say that I have raised my arm. By contrast, if my arm is lifted by a string or jerked upward by a spasm, the causal history of that movement disqualifies it as action.

Philosophers who have pursued this strategy have disagreed about what types of causes are necessary for action, but most accounts emphasize the role of particular psychological states – intentions, volitions, reasons, beliefs, desires, willings, tryings, and so on. These mental states are taken to be the efficient causes of active movements. Such *causal theories of action* can be found throughout the history of action theory. John Stuart Mill's discussion of action in *A System of Logic* clearly illustrates the sort of volitional account of action that held sway in eighteenth- and nineteenth-century philosophies of action:

> Now what is an action? Not one thing, but a series of two things; the state of mind called a volition, followed by an effect. The volition or intention to produce the effect, is one thing; the effect produced in consequence of the intention, is another thing; the two together constitute the action. I form the purpose of instantly moving my arm; that is a state of mind: my arm (not being tied or paralytic) moves in obedience to my purpose; that is a physical fact, consequent on a state of mind. The intention followed by the fact, or (if we prefer the expression) the fact when preceded by and caused by the intention, is called the action of moving my arm.[2]

Donald Davidson is often credited with the most famous contemporary forms of the causal theory of action. Davidson first argued that the cause of any action is the agent's primary reason for performing it, where a primary reason is defined as follows:

> R is a primary reason why an agent performed the action A under the description d only if R consists of a pro attitude of the agent towards actions with a certain property, and a belief of the agent that A, under the description d, has that property.[3]

To give a rationalizing explanation of this sort is to make sense of the activity under a description in light of the agent's beliefs and desires. For example, suppose that young Colin wants to make a mess and throws his oatmeal on the floor. This event can be given a rationalizing explanation in terms of Colin's mental states under the description 'making a mess': Colin wanted (a pro-attitude) to make a mess and believed that throwing his oatmeal would bring one about. By contrast, the event may have no rationalizing explanation under the description 'frustrating Colin's parents' because Colin may either have been unaware that throwing the oatmeal would frustrate his parents or had no such desire. In this case, we say that Colin *intentionally* makes a mess but *unintentionally* frustrates his parents, even though the mess-making event is identical with the parent-frustrating event.

On this account rationalizing explanations of human action are causal explanations: to explain *why* Colin threw his oatmeal is to explain *how* his oatmeal-throwing came about. Colin's behavior is intentional action under the description 'making a mess' in virtue of its being caused by a primary reason that makes reference to a future event under that very description. The belief and desire constitute Colin's primary reason for acting as he did. Had Colin's beliefs and desires been significantly different, he would have behaved differently.[4]

Davidson later revised his theory of intentional action to accommodate cases of 'pure intending' where an agent has intended to perform some action but has not yet embarked upon performing it. On his revised account, an agent A intentionally φ-s just in case A's φ-ing is caused by A's intention to φ, where an intention is an all-out evaluative judgment about the desirability of the intended action. The relevant point for our purposes is that on both accounts the status of a behavior as action depends upon the causal contributions of rationalizing propositional attitudes such as beliefs, desires, intentions, or judgments.

Davidson's view is far from the only popular contemporary intellectualist account of action. Michael Bratman rejects Davidson's accounts in favor of a functionalist view of intentions as 'conduct-controlling pro-attitudes' that serve as plan components and factor into practical reasoning as commitments and constraints.[5] These intentions also serve as reasons, although Bratman argues that they are 'framework reasons' which differ from the sorts of reasons associated with beliefs and desires in their broad structuring effects on belief/desire reasons, determining the relevance of those mental states for our ongoing practical reasoning.[6]

These theories share the presumption that a degree of cognitive sophistication is necessary for the production of action – a belief/desire pairing, an all-things-considered judgment with propositional content, or a plan component subject to rational norms. This comports with action theory's general emphasis on deliberation and planning, as exemplified by this quote from Bratman:

> [W]e are planning creatures. We form future-directed intentions as parts of larger plans, plans which play characteristic roles in coordination and ongoing practical reasoning; plans which allow us to extend the influence of present deliberation to the future. Intentions are, so to speak, the building blocks of such plans, and plans are intentions writ large.[7]

Clearly not all intentional actions are products of planning and deliberation – life would be exhausting if every act required conscious ratiocination.[8] But if Bratman and Davidson are right, action is necessarily produced by mental states that *can* enter into deliberative processes. This suggests that the capacity to act is closely related to the capacity for deliberate, future-directed planning, and that the possession (if not always the exercise) of these latter capacities is necessary for agency more generally.

A terminological remark may be helpful for negotiating what follows, as some readers unfamiliar with action theory may assume that the very status of an action as intentional logically entails the participation of an intention in its production. But although some philosophers have argued as much,[9] the matter is contentious. I use the term 'intentional' in a broad sense that does not presuppose any particular stance on this matter. Even if we were to discover that eliminativist theories of mind were true and that no brain state could be properly described as an intention, we would nonetheless be justified in describing certain types of behavior as intentional, or as having been performed intentionally.

This broader sense of the term 'intentional' allows us to coherently assert that one may act intentionally without forming an intention to act. To further muddy the waters, recent philosophical work has focused on 'nonintentional action', which is intentional in the ontologically uncommitted sense used here, but nonintentional in that it does not depend upon the presence of intentions, reasons, etc. In hopes of maintaining some semblance of clarity, I will at times refer to actions that depend upon intentions (or whatever states intentions might be reducible to) as 'full-blooded' or 'robustly' intentional actions. This chapter might be usefully described as an attempt to show that not all intentional action is full blooded.

It seems clear that, details aside, many of our actions do depend in various ways upon our beliefs, desires, intentions, and so on, or at least they are usefully described as so depending. However, I want to suggest that the traditional approach has missed something vitally important about action by focusing on sophisticated cases of human agency and that there is much to gain by adopting a broader view. There are behaviors that, although failing to meet Davidsonian standards for intentional action, are nonetheless genuine actions. These behaviors include (but are not limited to) two general classes of action:

P1. The purportedly 'mindless' activities of human agents, which do not depend on the presence of a motivating or guiding intention.
P2. The active movements of nonhuman animals that lack the capacity for deliberative thought.

We might describe these classes of behavior as forms of *primitive* action. P1 suggests that obvious agents like humans can express their agency in subintentional fashion. A small but growing number of philosophers have endorsed similar points, advancing accounts of 'automatic action',[10] 'arational action',[11] and 'nonintentional action'.[12] Bratman himself suggests the possibility of 'spontaneous action' which is voluntary and purposive but which requires no motivating intention.[13]

P2 suggests that systems incapable of planning and deliberation can nonetheless act. Harry Frankfurt offers the example of a spider's moving its leg as a case of nonhuman animal action. Although we may metaphorically describe the spider as intending to move, it is unlikely that the spider is capable of implementing anything like a future-directed intention, to say nothing of an all-things-considered judgment about the desirability of a given act.[14] Nonetheless, in offering even a figurative description of the spider's movements, it seems that we are picking out something closer to intentional action than we would if we were to describe a vending machine as intending to discharge soda, or a billiard ball as intending to roll when struck. The spider's movement is not *mere* movement, as it would be if its leg were blown by the wind or manipulated by a curious arachnologist. Frankfurt argues that traditional causal approaches to action ignore this fact. This failing leads us to exaggerate the uniqueness of human action and distorts our understanding both of the nature of action and of the human agent's place in the world:

> The conditions for attributing the guidance of bodily movements to a whole creature [. . .] evidently obtain outside of human life. Hence they cannot be satisfactorily understood by relying upon concepts which are inapplicable to spiders and their ilk. [. . .] [W]e must be careful that the ways in which we construe agency and define its nature do not conceal a parochial bias, which causes us to neglect the extent to which *the concept of human action is no more than a special case of another concept whose range is much wider.*[15]

Other philosophers have suggested that the difference between humans and even very simple nonhuman animals may not be as stark as we typically presume:

> In attempting to understand the elements out of which mental phenomena are compounded, it is of the greatest importance to remember that from the protozoa to man there is nowhere a very wide gap either in structure or in behaviour.[16]
>
> Most of the time we human beings may operate as agents in the unthinking manner of other animals.[17]

This broader concept of action is what I am calling 'primitive action'. The term was coined by Tyler Burge to describe basic forms of nonhuman animal action, but I extend the usage to cover both P1 and P2. This extension may not be warranted. It may be the case, for example, that humans are the only genuine agents, but can nonetheless exhibit a form of agency distinct from the full-blooded variety. Alternatively, it may be that all human action is full blooded but nonhuman animals possess a uniquely nonintentional sort of agency. However, as Frankfurt's quote suggests, the behaviors described in P1 and P2 may be related in interesting ways.

Indeed, it is plausible that there might be some continuity between the active behaviors of organisms simpler than ourselves and our behaviors in our less deliberative moments. Nature tends to create new capacities by modifying, combining, and leveraging more phylogenetically basic capacities in response to selection pressures. Perhaps our planning, deliberative, full-blooded agency might be better thought of not as a *sui generis* ability, but rather as an extension or special application of more basic adaptive capacities shared by a variety of physical systems.

This view is expressed vividly in a footnote about 'senseless action' by Brian O'Shaughnessy:

> Does nothing lie between a corpse-like graven image and a vehicle for reason? How *else* but as action is one to characterize the making of these movements,

and to what but *the person* is one to attribute them? One can hardly tele-
scope them into mere spasms [. . .] They relate to standard examples of action
somewhat as do objects that are mere lumps of stuff, say raw diamonds, to
objects that are mere lumps of stuff *and more*, e.g., artifacts, natural kinds
[. . .] Excluding them from the class of all actions would be roughly akin to
excluding gold nuggets from the class of material objects.[18]

O'Shaughnessy came to reject the concept of nonintentional action as incoherent, and even in the work referenced here he downplays its significance. But his
analogy with raw materials and refined objects is apt. If full-blooded intentional
agency is a refined form of a more basic sort of capacity, we might elucidate
our understanding of it by analyzing its raw materials, as well as acquire further
insight into our place in the world by seeing where else those materials might
happen to turn up.

To this end a focus on simpler organisms is helpful: by determining the precise level of complexity at which whole-organism behavior arises, we can iso-
late some of the core properties of agency and identify what additional materials
might be required. Upon reaching a satisfactory general account, we can then see
how primitive agency is elaborated or extended in more sophisticated systems,
to include agents like ourselves. This bottom-up approach allows us to avoid the
'parochial bias' Frankfurt cautions us against: in starting at the level of human
agency and working our way down, we may adopt the wrong criteria for classify-
ing behavior as action.

This approach is well supported by reflection on the tumultuous recent history of cognitive science. Twentieth-century cognitive psychology took deliberate, reflective thought to be the paradigm case of cognition and attempted to
extend that concept downward to explain less introspective activities. Although
early advances in language processing and formal reasoning were encourag-
ing, the problems with this approach were exposed by attempts to implement its
principles in autonomous robots. The resulting artificial systems, modeled after
Cartesian thinking things, could only engage the world through computationally
costly symbolic intermediaries. Real-time, real-world sensorimotor coordination
problems were beyond them. Many of the false starts in cognitive science and
artificial intelligence can be attributed to the distortion introduced by this view
of cognition.

Many thinkers have reacted to this result by initiating their investigation of cognition at the level of basic embodied agents operating in real-world environments
and scaling their accounts up to human thinkers by gradually adding complexity
to those basic systems. This 'bottom-up' approach to cognition has paid theoretical dividends. Although deliberation-centered computational theories of mind
might once have been 'the only game in town',[19] recent work in 'embodied' and
'enactive' cognitive science has yielded bold antirepresentationalist approaches
that purport to explain even higher-order deliberative processes as extensions of
the processes governing simple movements. The gambit behind this book is that
a similar perspectival shift will yield equally rich results for the study of action.

The properties of primitive action

O'Shaughnessy characterizes action as a *natural kind*, and in seeking a broad notion of action that applies to a variety of active movements, it will be helpful to think of action as such. Unsurprisingly, the varied instances of this broad kind will not lend themselves to a uniform characterization. I adopt a Boydian account of natural kinds as homeostatic property clusters – families of contingent properties where some subset of those properties (or an underlying mechanism) serves to maintain the others. Natural-kind terms refer to objects that instantiate *most or all* of these properties. Here I attempt a characterization of the core properties of action and outline the dependence relations that hold between them. Action can be viewed as a cluster of six related properties:

> A1. Actions are behaviors produced by agents, not by mere parts of agents or external forces.
> A2. Actions are coordinated behaviors.
> A3. We can intervene to stop our basic actions.
> A4. We have a distinctly nonobservational sense of agency when we act.
> A5. In acting, we are unsurprised to observe that we have acted and uniquely surprised when our actions are thwarted.
> A6. Action has success and failure conditions.

On the theory of natural kinds adopted here, it is a contingent matter as to which of a kind's noncore properties are instantiated in a given instance of the kind. I take properties A1, A2, A3, and A6 to be essential to the kind. A4 and A5, by contrast, may be obtained only in agents sophisticated enough to token metacognitive states. But our sense of agency is a powerful (albeit fallible) indicator of when our behavior instantiates the core properties of action, and we will see in this chapter that reflection on the phenomenology of human agency can elucidate our concept of agency more generally.

A1. Actions are behaviors produced by agents, not by mere parts of agents or external forces

To endorse A1 is not to endorse anything like agent causation, but only to suggest that insofar as anything acts, it is the agent that acts, not its parts. *I* eat; *my stomach* digests. *Marisa* walks to the library, but *her legs* do not, and this is true even though her legs contribute essentially to the successful production of the action. We consider many movements of nonhuman animals to be 'whole-organism behavior' as well. We find sentences like the following incoherent:

> *The cat chased the dog, and so did its head, torso, legs, and tail.*

A cat can be coherently said to chase animals, but its constituent parts cannot, and this is so even though those parts compose the cat. The same can be said for other active behaviors: eating, hiding, mating, burrowing, and so on.

Admittedly, the bare appeal to ascriptive practice can only take us so far because some nonactions are attributed to agents. We can say 'John shivered' or 'Mary sneezed', but we do not mean to suggest that shivers and sneezes are actions.[20] In the case of shivering we still have a linguistic indicator of nonaction: although we might say that John shivered, we would be equally warranted in saying that John's body shivered. But this does not hold for other involuntary motions: Mary sneezes; her respiratory system does not. Things are complicated further by certain literary tropes – 'Graham's legs carried him across the dining room' – that ascribe actions to parts of agents.

Artful metaphors aside, we might note that the case of Mary's sneezing might be explained by the degree of control that Mary exerts over her behavior. Although her sneeze is in some sense involuntary, Mary does have some control over where she sneezes (she could leave the room, feeling a sneeze coming on) or how she sneezes (perhaps choosing to cover her nose and mouth, or stifling the sneeze so as not to disturb others). Defecation is a similar borderline case. In these cases we may ultimately attribute the movements to agents because we recognize them as partial expressions of agency; the agent retains some control over how they behave in these cases.

Of course, it remains to be seen what it is in virtue of which we correctly attribute Marisa's library-directed walking to Marisa rather than to her legs, her motor cortex, or some other collection of her parts. I attempt an answer in Chapter 4. But for now our intuitions that we capture something significant about agents in attributing actions to them rather than to their parts can help to orient our discussion.

A2. Actions are coordinated behaviors

To call a behavior *coordinated* is to say that its successful performance depends upon the integration of information from some combination of the environment and the various contributing subsystems of the agent. The successful performance of even our most basic physical actions requires the integration of multiple subsystems: motor, respiratory, visual, tactile, proprioceptive, etc., in such a way as to bring about a stable, environment-responsive pattern of behavior. The character of the behavioral product is determined partially by the individual contributions of these coordinated systems: one walks differently when out of breath, for example.

By contrast, many nonactive movements can be explained by appeal to the function of an isolated mechanism. The knee-jerk reflex can be explained by appeal to a simple reflex arc. The timing, direction, and force of the resultant kick are mere functions of external perturbation. We say that the stomach digests because this is a function of the stomach: there is some single entity doing the digesting. But when we say that Colin plays basketball, we do not mean that Colin is a single mechanism performing a basketball-playing function. Colin, like any human, is a remarkably complex bundle of subsystems, and his action results from the coordinated contributions of those subsystems.

Coordination of some sort, then, seems essential for agency. Indeed, the deep interdependence of perception and action has led some philosophers to suggest that agency appears precisely at the point where a degree of sensorimotor integration obtains within a system. O'Shaughnessy claims that newborn infants are agents precisely for this reason:

> The infant's waking on birth is his coming into an inheritance wherein some measure of integration obtains between at least some of his motor and tactile powers. That is, into a measure of ego development.[21]

This notion of ego development suggests that agency is closely associated with what some have called the sensorimotor self: a 'synthesizing mental centre' that serves as a locus for information integration and distribution. Similarly, Ulric Neisser's 'ecological self' emerges as an essentially embedded entity that is perpetually engaged in sensorimotor information integration and activity relative to a particular environment.[22]

The relationship between A1 and A2 is complex. We might say that the phenomenon described in A1 is prior to that described in A2, in that coordination requires a coordinating faculty of some sort, and indeed I think that the agent's primary role is to coordinate its various submechanisms (both with each other and with the agent's environment) in adaptively successful ways. But philosophical and psychological examinations of selfhood have been hard-pressed to identify a single mechanism that could be identified uniquely as a coordinating faculty. Indeed, we do not seem to find *any* agency in a system until the proper sensorimotor coordination obtains. We would not say of a dismembered organism that it was still an agent that happened not to be doing any coordinating at that moment. Agency depends on coordination.

Thus the phenomena described in A1 and A2 seem to be deeply interdependent: an agent is a whole that coordinates the operations of the very parts on whose coordination the agent depends! I will examine this relationship in later chapters, but readers familiar with complexity theory may have an idea of what is to come. For now it will suffice to say that A1 and A2 are core properties of any sort of action.

A3. We can intervene to stop our basic actions

As the authors of our coordinated movements, we possess the capacity to intervene and stop their production.[23] If I am walking, I can stop myself from walking. If my behavior is being controlled remotely, I likely cannot. The epileptic cannot simply halt his seizures, nor can the Parkinson's sufferer intervene to stop her trembling. Seizures and tremblings befall their sufferers; they are not things that agents do.

This capacity is most evident in the performance of actions that depend entirely on the agent's ability to intervene in the unfolding of an action, such as willfully coasting or gliding. A driver who allows her car to coast downhill guides its motion even though she makes no direct causal contribution to the coasting. This is so because the driver is positioned to intervene in its unfolding if necessary. This positioning can be understood, again, in terms of an agent's coordinating her behaviors with an environment – even passive guidance requires coordination of perceptual and motor subsystems to allow the agent to monitor his action and intervene in a timely fashion when necessary, adjusting behavior in response to changes in environment. If this coordination is lost, the agent has lost control of her action; it is no longer her own. Thus A3 depends upon A1 and A2.

A4. We have a distinctly nonobservational sense of agency when we act

A4 and A5 are contingent properties of action and may not arise in agents that satisfy A1–A3 but lack the additional metacognitive machinery required for self-knowledge. But it is worth discussing the character of our awareness of action at some length here to show how it differs from other forms of awareness. We will see that the agent's distinct awareness of his bodily movements depends essentially upon the core properties of action discussed thus far.

G.E.M. Anscombe categorizes intentional action within the class of 'movements known without observation'.[24] She takes knowledge of limb position to be paradigmatically nonobservational: for example, the knowledge that one's legs are bent does not require a characteristic sort of tingle in the knees or any sort of visual examination. Rather, proprioceptive awareness suffices for knowledge of one's body position.

An extreme deficit of this awareness is exhibited in somatoparaphrenia, or *alien hand syndrome*. The somatoparaphrenic retains feeling, control, and perceptual awareness of his limbs, but suffers from a perceived loss of ownership over them. He views the alien limb as belonging to others or as being 'possessed'. To give one account:

> [This arm] is not mine. I found it in the bathroom when I fell. It's not mine because it's too heavy; it should be yours. I can move and do everything; when I feel it too heavy, I put it on my stomach. It doesn't hurt me; it's kind.[25]

The somatoparaphrenic views his limb entirely from the standpoint of an observer. This suggests that proprioceptive awareness of one's limb position is distinct in kind from the sort of awareness that one achieves through observation alone.

The epistemic relationship between agents and their actions is similarly distinct from that of other forms of observational knowledge. Anscombe gives the example of a window-opening agent who is asked about her behavior:

> I do open the window; and that means that the window is getting opened by the movements of the body out of whose mouth those words come. But I don't say the words like this: 'Let me see, what is this body bringing about? Ah yes! The opening of the window!'[26]

Agents require no evidence of their actions beyond the fact that they are acting. We know what we are doing in doing it, at least as far as our basic actions are concerned.[27] Anscombe distinguishes this 'agent's knowledge' from 'the contemplative model', which she takes to encompass standard cases of observational knowledge. She likens the agent's knowledge to the knowledge that an architect has of a building that he is designing.[28] If the architect knows that his workers are reliable, he does not need to visit the worksite to know what sort of building he is creating. He knows this because he is giving the orders, and in giving them he is aware of the building's construction in a way that a passing observer of the worksite is not.[29]

This suggests that the structure of 'agent's knowledge' is fundamentally distinct in kind from exteroceptive observation. O'Shaughnessy describes it as 'at once intuitional *and* immediate *and* almost invariably attentively recessive *and* such that both the target-zone and goal-event are given without mediation by concepts'.[30] Indeed, we cannot take the observer's perspective toward our own actions while performing them, as such a task would cause 'an impossible divide in experiential consciousness'.[31] The impossibility of this task is not a matter of cognitive/attentive resources, but rather results from the fact that the relation an agent bears towards her action is fundamentally distinct in kind from the relation an observer bears toward that action. O'Shaughnessy argues that taking both perspectives would require an impossible 'fragmentation of the self'.[32]

To make sense of this fragmentation, consider the interdependence of perception and action. Gibsonian perceptual psychology suggests that perception involves the registration of affordances for action: one perceives his environment directly as permitting various forms of interaction.[33] Thus the nature of perception depends in part upon one's capacities for action. Moreover, the successful production of an action depends upon informational feedback. In the case of raising one's arm, the production of the action requires, at a minimum, continuous proprioceptive feedback. This feedback is necessary to the task of coordinating one's actions (per A2).

For example, in raising my arm I receive feedback that includes the information that it is *my* arm that I am raising and that the position of my arm at that time is such that I could divert its course in various ways (or halt it altogether, as per A3), etc. My perception of the acts of others, by contrast, includes no such information: I do not view your arm-raising as something that I can directly control. I view it from the standpoint of an observer rather than that of an agent. O'Shaughnessy's point is that in attempting to view one's own action from a third-personal perspective (i.e., from the stance of a 'pure' observer), one attempts to devote attention to two fundamentally distinct kinds of cognitive task. Indeed, O'Shaughnessy goes so far as to argue that it is impossible to view the action from both standpoints: insofar as we view our own action from the standpoint of an observer, we cannot view it from the standpoint of an agent, because to view one's action in this way *just is* to relinquish the very sense of agency that is associated with the first-personal character of action:

> To succeed in the question of self-observation, would in effect be to succeed in becoming two people, the being who acts and the being who observes that action, such that the former has to play the role of *object* for the latter![34]

We perceive the world as affording opportunities for interaction. We perceive our own actions as processes we could halt or modify during the course of their

production. We perceive the actions of others differently. We may be able to intervene in the production of others' actions, but only by obstructing them or by
persuading their agents to adjust them in various ways. We cannot intervene in
the acts of others as their agents can, and we do not, as observers, perceive those
acts as suitable candidates for such intervention. Thus, perceiving an act from the
dual standpoint of agent and observer would require us to perceive it as a process
in which we both can and cannot intervene. This cognitive dissonance may be the
source of what O'Shaughnessy refers to as the 'fragmentation of the self'.[35]

In discussing our sense of agency, I have said much about proprioceptive
awareness, but although proprioceptive feedback is essential for agency, it is not
sufficient for our sense of agency. We can see this by considering the case of
anarchic hand syndrome (AHS). Whereas alien limb syndrome is associated with
the loss of a sense of *ownership* of a limb, anarchic hand syndrome is characterized by a loss of control of limbs, which seem – both to outside observers and to
the sufferers themselves – to behave autonomously. To know what her hand is
doing, the sufferer must watch it in the same way that she must watch others. As
Marchetti and Della Sala write:

> One of our patients (GP) once, at dinner, much to her dismay saw her left hand
> taking some fish bones from leftovers and putting them into her mouth . . . .
> Another patient of ours (GC) often complained that her hand did what it
> wanted to do, and tried to control its wayward behaviour by hitting it vio-
> lently or talking to it in anger and frustration.[36]

GC describes her hand as doing what *it* wants to do, not because the hand is actually intelligent, but because its movements (which, the authors note, can be goal
directed) are not *her* actions. This is true even though the agent has full perceptual
awareness – including proprioceptive awareness – of the anarchic hand's movement.
Thus proprioceptive awareness alone does not suffice for our sense of agency. The
missing component is a sense of control; the loop between perception and action has
been severed, and the agent cannot successfully coordinate her movements.

Thus we can now see how A4 depends not merely on feedback, but also on the
sum of processes associated with sensorimotor coordination. Our sense of agency
depends essentially on the core properties of action, none of which are satisfied
in cases of AHS: The sufferer as a whole (A1) cannot successfully coordinate his
behavior (A2), as evidenced by his inability to voluntarily stop the hand from
moving (A3).[37] The patient loses agent's knowledge of his movement *because* he
is no longer controlling it.

A5. In acting, we are unsurprised to observe that we have acted and uniquely surprised when our actions are thwarted

Wittgenstein remarked that 'voluntary movement is marked by the absence of surprise'.[38] I take this lack of surprise to be the direct result of A4: the nonobservational awareness that amounts to our sense of agency. In acting I am immediately

aware of my action, and thus when I move my hand I am unsurprised that my hand moves. Some philosophers have suggested that this lack of surprise is the result of a future-directed intention: if I form the prior intention to move, then I am unsurprised when the event described in the content of that intention takes place. But this treats the phenomenon as something akin to a case of fulfilled expectation. When I throw a ball to my dog (who reliably fetches), I do so with the expectation that he will return the ball. I am unsurprised when he does so, and this lack of surprise results directly from the satisfaction of my expectation about my dog's behavior. But this is different from the sort of relationship I have to my own action. In the case where my dog fails to return the ball I may be surprised, but nothing more. By contrast, if I attempt to move my arm but cannot, I am surprised in a quite different way: there is something phenomenologically unsettling about one's basic actions being thwarted that is not present in other cases of mere unsatisfied expectation.

Nor is it the case that a lack of surprise about some state of affairs generally entails the existence of a future-directed mental state with content describing that state of affairs. For example, I am used to the sound of birds chirping when I leave my house in the morning. Upon opening my door, I hear a bird chirp. I am unsurprised to hear the sound, but this is not to say that I planned, intended, or even expected to hear it. Rather, bird chirping is an ordinary feature of my environment to which I have fully habituated. Similarly, I am unsurprised when I act, not because I have fulfilled some expectation that I will move in an intended way, but rather because it is part of my ordinary life as an agent that when I try to move about in certain ways I meet no resistance. I am surprised only when I do meet such resistance. If an impediment to my acting was to become a pervasive feature of my life (as in the case of paralysis), I would eventually habituate and be entirely unsurprised by my limbs' new way of moving. Thus I suggest that the lack of surprise that accompanies our actions results not from a prior mental state (which may, of course, be active in the production of certain types of action), but rather from the sense of agency described in the discussion of A4.

However, when my action is impeded in a way to which I have not habituated, there is a jarring quality to the experience that is not present when an expectation about the behavior of others has been thwarted. Consider the act of throwing a ball to someone: In throwing, I control the trajectory of my arm in such a way as to guarantee that the ball pursues a particular path terminating at the receiver. My focus is on the receiver, not on my arm. If my arm is grabbed by someone else before I release the ball, however, I am immediately aware that something has gone wrong. I have lost control of my movement in a certain way, and this sense of lost agency immediately draws my attention away from the receiver and toward the new obstacle.

Why have I devoted so much attention to A4 and A5? One reason is that the other, more fundamental properties will occupy most of the rest of the book. Another is that our sense of agency is sometimes taken to show that action can only be understood on an intellectualist framework, as though the characteristic awareness of our action could only be explained by appeal to the participation of the propositional attitudes. But I have argued that the sense of agency arises from successful agent-driven sensorimotor coordination, not from beliefs or intentions. In the coming chapters I will examine how this sort of coordination could obtain in their absence. This is not to say that the propositional attitudes have no place in a complete action theory (they do, in the full-blooded variety of action), but rather that the core properties of action understood broadly do not require their participation.

A6. Action has success and failure conditions

Our most basic actions are subject to norms that mere happenings are not. If a dislodged rock happens to roll down a hill, it has neither succeeded nor failed at reaching the bottom. The thrashings that accompany a seizure are unfortunate, but they are neither successes nor failures. By contrast, when I begin to raise my arm, I am performing an action that can fail. Perhaps its course will be impeded, or just at that moment I will suffer a stroke that causes my arm to fall limp at my side. When all goes well, the action is a success, and most of our movements are successful. Sitting up, walking, hand raising: the unambitious among us can take credit for innumerable tiny successes over the course of each day.

The normative character of action is essential for the individuation of equivalence classes of action-kinds within the broader kind of action. Consider the act of feeding. While I eat my meal, the housefly next to me eats its meal. Houseflies eat solid food by vomiting their stomach contents onto it, breaking it down for easy consumption by proboscis. My methods are considerably neater. Given the very different ways in which we feed, what similarities would establish that we are both feeding? The normative character of the action – what we are performing these behaviors *for*, which in turn would involve specification of the conditions for their successful performance – allows us to taxonomize mechanically disparate actions on the basis of functional similarities.

This property of action, like the others, depends on A1 – the fact that agents cause actions. This is due to the fact that success and failure conditions must be assigned relative to some standard. In the cases of thwarted action mentioned thus far, the problem has arisen when the agent fails to properly coordinate his actions either internally (as in the case of my sudden stroke) or with the environment (when my hand is unexpectedly restrained mid-throw). We might think of these as cases of failed coordination, but what distinguishes a case of failed coordination from a successful case?

Failure and success are not absolute concepts. Success is success *for* something – in this case, for the agent. In acting, the agent determines the success conditions for its action. In cases of full-blooded intentional action, the standards for success are (nearly) settled by the content of the motivating intentions: if I act on an intention to act in some way, then I have failed if I do not so act. But we will see in Chapter 5 that the normative character of action can be determined in other ways, and in fact extends directly from the most basic organizational principles of the living organism.

The idea that norms are specified relative to kinds of systems will also help us identify cases when A1 is satisfied, that is, cases where the agent, rather than one of its parts, is acting. Later in this book I offer a general strategy for ascribing behavior to a given system as such. This involves (1) identifying a domain of inter-actions obtaining at the operational boundary between a system and its environ-ment, (2) specifying the norms governing those interactions, and then (3) ascribing behavior to that system only when that behavior is governed by the appropriate system-level norms. I will argue that there are distinctly *agential* norms of behav-ior and that behavior is properly attributed to the agent as a whole just when its generation is governed by agential norms. I offer this argument in Chapter 6.

Thus far I have attempted to motivate the project of seeking a broader form of agency than is traditionally examined in the philosophy of action, and I have sketched its core properties. Action in this broad sense is defined at its core by whole-agent behavior and coordinated activity. The rest of this book is an attempt to flesh out the details of this picture. How do we make sense of the codependence of A1 and A2? What does coordination of the right sort require? How do systems incapable of forming full-blooded intentions establish proper norms of action? What distinguishes a primitive agent from its parts, or from other collections of organized matter? And why should we think that analysis of these primitive sys-tems sheds any light on what *we* do? In what follows I attempt to answer these questions, starting with the notion of coordination in Chapter 2.

Notes

1 L. Wittgenstein, *Philosophical Investigations* (Malden, MA: Blackwell Publishing, 1958), §621.
2 J.S. Mill, *A System of Logic* (London: John Murray Ltd., 1961), p. 35.
3 D. Davidson, 'Actions, Reasons and Causes', *The Journal of Philosophy* 60:23 (1963), pp. 685–700, on p. 687.
4 Strictly speaking, beliefs and desires cannot be causes on the Davidsonian picture. This is because for Davidson, causes are events, but beliefs and pro-attitudes are not events but rather states. Thus the true cause of action is not the belief, but the 'onset' of the belief. See E.J. Lowe, 'Self, Agency and Mental Causation', *Journal of Consciousness Studies* 6, pp. 225–239, on p. 225.
5 M. Bratman, *Intention, Plans and Practical Reason* (Stanford, CA: CSLI Publications, 1999), p. 27.
6 Ibid., p. 34.
7 Ibid., p. 8.
8 The issue is obscured by our ways of speaking about action. For example, I might perform an action deliberately without actually engaging in any deliberation prior to performing it. In this sense of 'deliberately' I mean simply that I perform the action intentionally. In these cases we might say that I performed the action *on purpose*, where the purpose can be specified by my intentions.
9 Bratman (1999) calls this 'the simple view' of action (p. 112).
10 E. Di Nucci, 'Automatic Actions: Challenging Causalism', *Rationality Markets and Morals* 2:1 (2011), pp. 179–200.
11 R. Hursthouse, 'Arational Actions', *The Journal of Philosophy* 88:2, (1991), pp. 57–68.
12 D.K. Chan, 'Non-intentional Actions', *American Philosophical Quarterly* 32:2, (1995), pp. 139–151.

13 Bratman, *Intention, Plans and Practical Reason*, p. 126.
14 This is certainly true if such implementation depends upon a capacity for language use.
15 H. Frankfurt, 'The Problem of Action', *American Philosophical Quarterly* 15 (1978), pp. 157–162, on p. 162. My emphasis.
16 B. Russell, *The Analysis of Mind* (New York: Cosimo, Inc., 2004).
17 P. Pettit, 'The Reality of Group Agents', in C. Mantzavinos (Ed.), *Philosophy of the Social Sciences: Philosophical Theory and Scientific Practice* (Cambridge, UK: Cambridge University Press, 2009), p. 72.
18 B. O'Shaughnessy, *The Will: A Dual Aspect Theory* (2nd ed., Cambridge: UK: Cambridge University Press, 2008), pp. 54–55 note 2 (original emphasis).
19 J. Fodor, *The Language of Thought* (Cambridge: Harvard University Press, 1975).
20 Burge (2009) suggests that we might attribute these behaviors to agents rather than their parts because the actions manifest as whole-body movements and thus have the appearance of coordinated animal behavior.
21 B. O'Shaughnessy, 'Trying (as the Mental 'Pineal Gland')', *The Journal of Philosophy* 70:3 (1973), pp. 365–386, on p. 324.
22 U. Neisser (ed.), *The Perceived Self: Ecological and Interpersonal Sources of Self-Knowledge* (Cambridge, UK: Cambridge University Press, 1993), p. 4.
23 This criterion is restricted to basic actions because intervention is frequently impossible in the case of nonbasic actions. For example, we can still say that Oswald shot JFK even though he couldn't have intervened to stop the bullet after pulling the trigger. By contrast, he could have stopped himself from pulling the trigger. It should be stressed that this is one virtue of treating action as a homeostatic property cluster – a movement needs only to satisfy *most* of these criteria to count as action.
24 G.E.M. Anscombe, *Intention* (Oxford, England: Basil Blackwell, 1957), p. 14.
25 G. Rode, N. Charles, M.T. Perenin, A. Vighetto, M. Trillet, and G. Aimard, 'Partial remission of hemiplegia and somatoparaphrenia through vestibular stimulation in a case of unilateral neglect', *Cortex* 28:2 (1992), pp. 203–208, on p. 206.
26 Anscombe, *Intention*, p. 51.
27 We may, of course, be ignorant of the various implications of our actions: for instance, I may be unaware that my hand waving is annoying the person behind me or insulting a person from another culture for whom that particular gesture is offensive. Knowledge of such implications would require observation that knowledge of one's own movements does not.
28 Anscombe, *Intention*, pp. 82, 88–89.
29 This is not to say that a purely nonobservational awareness of one's actions is sufficient for their *successful* performance. Anscombe recognizes that the successful performance of most action depends upon perceptual feedback: a blind person may intentionally write his name and know what he is doing without perceptual feedback, but without the capacity to perceive the boundaries of the paper or whether his pen has ink, he is less likely to accomplish his intended goal. But for Anscombe the knowledge that one is acting in a particular way does not depend upon perception – the role of perception merely serves as 'an aid' for the successful completion of the action (p. 53).
30 O'Shaughnessy, *The Will: A Dual Aspect Theory*, p. 313 (original emphasis).
31 Ibid., p. 339.
32 Ibid., p. 345.
33 J. Gibson, *The Ecological Approach to Visual Perception* (Boston, MA: Houghton-Mifflin, 1979).
34 O'Shaughnessy, *The Will: A Dual Aspect Theory*, p. 339 (original emphasis).
35 This point can be further supported by appeal to the commonplace fact that taking a third-personal, 'purely' observational perspective toward one's actions destabilizes those actions. It is difficult to catch a ball or drive a car when attempting to watch oneself performing the relevant action. There is a sense in which agents are *present* in their actions – they enjoy a mode of immediate epistemic access to what they are doing

that is distinct from the access that they have to other features of the world. Attempting to access one's actions both from the perspective of the agent and from that of the observer leads to confusion. The first-personal access that agents have to their own actions might be thought of as our *sense of agency*.

36 C. Marchetti and S. Della Sala, 'Disentangling the Alien and Anarchic Hand', *Cognitive Neuropsychiatry* 3:3 (1998), pp. 191–207, on p. 196.

37 It might be argued that in cases of AHS we are dealing with not one but two distinct agents – an actual fragmentation of the self! But even if this is so, the movements of the anarchic hand are not properly attributable to whichever agent is reporting its wayward behavior.

38 Wittgenstein, *Philosophical Investigations*, §628.

2 Guidance and deviance

Frankfurt on the causal theory of action

Thus far I have argued that the core characteristic of action, understood broadly, is it's being coordinated or guided by an agent. This conclusion will satisfy no one without a detailed characterization both of what the relevant sort of guidance could be and what it would mean for an agent to guide an action. This chapter is devoted wholly to the first question. What we will see is that the necessary sort of guidance for a genuine action (as opposed to a chain of movements that happens to involve an agent) requires a unique form of *synchronic* guidance that allows no spatiotemporal gap between the agent's guiding and its guided movement. In what follows I motivate this requirement and show how difficult it has been for even philosophers aware of the problem to satisfy it. I then offer my own attempt in Chapters 3 through 5.

It will be easiest to see what the necessary sort of guidance could be by contrasting it with what it cannot be. Consider this simplified story of an arm raising: At time T1, I form the intention to raise my arm at time T2. Then, at T2, my intention causes my arm to rise. We might view what has happened as a form of guidance: my intention at T1 initiated a process that included my various parts' coordinating to produce a coherent movement that satisfied the content of that intention. Had I not formed this intention, I might have performed some other movement, or done nothing at all. We might be comfortable describing many similar processes as forms of guidance. The author of a list of chores might be thought of as guiding the behavior of an intended recipient, assuming that all goes well and the content of the list is satisfied. Similarly, intentions might be thought of as action guiding in the sense that they function as submitted requests or commands to the body.

The problem with such approaches to guidance is that they restrict the agent's role to that of supplying the right sort of antecedent cause. In doing so they impose a problematic gulf between the motivating states of the agent and the actions that they are purported to guide. Frankfurt's critiques of causal theories of action are clearly motivated by this problem. Frankfurt argues that causal theories (1) entail that actions are indistinguishable from mere happenings at the time of their performance, (2) entail that we cannot know that an agent is acting unless we know certain relevant facts about their history, and (3) are subject to deviant causal chain counterexamples. He rejects the idea that the core features of action could be adequately characterized in terms of a movement's causal history, arguing that action requires the ongoing guidance of the agent throughout its performance.[1]

Frankfurt argues for (1) on the grounds that if actions and mere movements differ only in their causal antecedents, it follows that the movements associated with action and the movements associated with mere movement will be identical, differing only with respect to their causal histories. The causalist answers Wittgenstein's question – what is the difference between my raising my arm and my arm's merely going up? – by appealing to the presence or absence of a prior motivating mental state. But this means that once the action has been determined, the agent plays no further role in its production. This picture of action is impoverished. When I raise my arm, I am, within certain limits, able to make various adjustments to my action throughout its performance (importantly, *I* am able to make these adjustments). This continuous process of monitoring and adjustment is essential to guidance as Frankfurt conceives of it. We may contrast this sort of movement with ballistic movements such as kicking, which cannot be modified once initiated. If the role of the motivating state is confined to that of initiating action, then *all* actions are ballistic from the agent's perspective – the prior intention to raise one's arm might cause the action, but we cannot think of it as guiding it. No motivating state can guide movement from the past.[2]

(2) is the epistemic upshot of (1). If the movements associated with action are indistinguishable from those associated with mere movement, then the only way we could know that an agent was acting would be to observe the agent prior to the initiation of the movement to see whether the relevant agency-conferring mental state has been tokened. But this result is absurd, as we can reliably classify movements as action when we observe them mid-performance. Moreover, it is unclear precisely what sort of evidence would count as justifying our belief that someone had tokened the appropriate intention *other than* the intentional movement itself. Frankfurt suggests that a given pattern of movement must include some ongoing indicator of its status as an action. Actions show us 'on their faces' that they are under the guidance of an agent.[3] This suggests that certain patterns of behavior indicate that their agents exert continuous control over them.

Deviant causal chain objections – (3) – are particularly troublesome for causal theories of action. Such objections take the form of cases where the allegedly necessary and sufficient conditions for action are met but the resulting movement is not action. To use Davidson's well-known example:

> A climber might want to rid himself of the weight and danger of holding another man on a rope, and he might know that by loosening his hold on the rope he could rid himself of the weight and danger. This belief and want might so unnerve him as to cause him to loosen his hold, and yet it might be the case that he never chose to loosen his hold, nor did he do it intentionally.[4]

In such examples the agent is no longer involved in the right way in the succession of consequent events. Even if those events are caused by his intention and satisfy its content, he nonetheless fails to act. Frankfurt argues that the problem in such cases is that the agent has lost control of the progression of events in such a way that they can no longer be viewed properly as her actions. This suggests the need for an additional sufficient condition for action that would guarantee this control. Frankfurt intends for guidance to do precisely this work.

Primitive action as guided movement

Frankfurtian guidance is an attractive starting point for an investigation of primitive action because it presupposes no sophisticated conceptual abilities on the part of the guiding agent. If we take the distinction between action and mere movement to hang on the nature of a prior cause and we recognize that at least many of our actions have intentions as their prior (contributing) causes, it is unsurprising that we would place deliberative reasoning at the core of our action theory. But although intentions may frequently guide actions, nothing about the guidance relation entails the presence of an intention. It may be the case that many organisms that cannot form explicit intentions may nonetheless guide their movements. This shift in focus from motivating states to the nature of guidance opens the door to a broader examination of agency.

Frankfurt offers a hierarchical theory of action that identifies purposive movement as the most fundamental form of behavior and full-blooded intentional action as its most sophisticated extension. Intentional movement, which Frankfurt identifies as the most basic form of action, occupies the middle ground. Any guided or directed movement counts as purposive movement, to include reflexive movements such as pupil dilation (here the guidance comes from the mechanisms that adjust pupil size in response to light, arousal, etc.). We might call behaviors 'merely purposive' when the relevant purposes are simply those that factor into explanations of subpersonal mechanism function. For example, we frequently blink without exerting direct control over our eyes to do so. In such cases, the blinking is explained by appealing to its primary function of cleansing and moistening the cornea. Thus blinking is purposive, although the relevant purposes are nothing like goals (occurrent or not) of the organism.[5]

Intentional movements – the most basic Frankfurtian actions – are cases of purposive movement where the requisite guidance is provided by the agent rather than by some subsystem therein. When my knee is tapped with a hammer, I do not take ownership of the resulting reflexive kick. The tapping activates a subpersonal perceptuomotor system – my leg moves but *I* don't move it. By contrast, in walking across a cluttered room I am in control of my various contributing subsystems, coordinating them in such a way as to avoid stepping on objects, making on-the-fly adjustments in the face of new obstacles, and so on. This is true even if I did not form the conscious intention to navigate the room. For example, I frequently pace around my office while working through my schedule for the day. In this case there is no appealing to my beliefs and desires in rationalizing my action, but the action is nonetheless mine in a way that the knee-jerk reflex is not.

Frankfurt defines intentional action as a subclass of intentional movement. Intentional actions are intentional movements 'undertaken more or less deliberately or self-consciously'.[6] Thus intentional movement is conceptually prior to intentional action: the performance of an intentional action requires an explicit intention that certain intentional movements should occur. But Frankfurt is clear that there is 'nothing in the notion of an intentional movement which implies that its occurrence must be intended by the agent, either by way or by forethought or by way of self-conscious assent'.[7]

Frankfurt's hierarchy can be summarized as follows:

1 Purposive movement: Guided movement of any sort.
2 Intentional movement (action): Purposive movement guided by the agent.
3 Intentional action: Intentional movements that are undertaken deliberately or self-consciously.

Frankfurtian action has its roots in basic purposive mechanisms, and intentional action is a narrow subclass of a more general notion of action that is to be understood in terms of the recruitment and guidance of purposive movements by the agent. This broader notion of intentional movement is applicable to human and nonhuman animals alike: purposive movement is ubiquitous in the biological world, and the notion of guidance does not presuppose capacities that would exclude nonhuman animals. As Frankfurt notes:

> The conditions for attributing the guidance of bodily movements to a whole creature, rather than only to some local mechanism within a creature, evidently obtain outside of human life. Hence they cannot be satisfactorily understood by relying upon concepts which are inapplicable to spiders and their ilk.[8]

Thus Frankfurt supplies the blueprints for a theory of primitive action as agent-guided movement that casts it as the 'raw material' out of which more sophisticated action is constructed. The remainder of this chapter is dedicated to understanding what the requisite form of guidance could be.

Nanodeviance and the agent/action gap

Frankfurt's work has demonstrated the necessity for guidance in action, but more must be done to establish what guidance amounts to or why we ought to abandon, rather than modify, traditional approaches to action. A defender of the causal theory of action could argue that intentions are plausible behavior-guiding states once we take the role of perceptual feedback into account. Once my intention to raise my arm initiates my behavior, I monitor my action to determine whether it will satisfy that intention, making adjustments as needed. Thus the guidance requirement seems at best a modest amendment to the traditional approach.

I suspect that the guidance requirement is not so easily met once we have a clear understanding of what it truly requires. First, I have already discussed at some length that our knowledge of our own actions seems to be distinct in kind from perceptual knowledge, so it is unclear that talk of monitoring or observation is useful. But a deeper problem is unearthed by Frankfurt's attack on causalism. I will show that taking deviance seriously forces us to adopt a highly specific form of guidance that cannot be explained in terms of mere observation and adjustment. This form of guidance constitutes the raw material out of which all other forms of action are formed.

Deviant causal chain counterexamples take three forms: primary, secondary, and tertiary. Primary deviance counterexamples insert some disqualifying event between the connection between the motivating states of the agent and the movement they produce, as in Davidson's story of the climber. Secondary deviance inserts itself into the causal process between bodily movements and consequences.[9] Tertiary deviance, or 'waywardness', involves processes antecedent to intention formation, introducing a disqualifying amount of luck into the formation of an action plan that is then successfully carried out.[10] In all such cases the agent loses control of her behavior in some significant way – her behavior is caused by her intentions, but the presence of deviance in the chain of events entails that her behavior is not up to her.

Deviant causal chain counterexamples are not unique to causal theories of action. The problem of deviance arises anywhere a theory of action allows a spatiotemporal gap between motivating states of the agent and the movements it produces. The demand to bridge the gap between agent and action may be expressed in terms of a condition of adequacy on theories of action:

> The proximity condition (PC): There must be no spatiotemporal gap between the motivating states of an agent and its action.

Most attempts to deal with deviant causal chain counterexamples involve the introduction of some element to satisfy PC. Frankfurt's appeal to guidance is one such attempt, and I will address others shortly. However, if a deviant causal chain counterexample can be constructed that satisfies one of these proposed proximity measures, it follows that the corresponding theory is inadequate. In what follows I will argue that each of the theories considered here fails to satisfy PC, and that indeed any theory that treats the relationship between agent and action as diachronic will fail to do so.

The case I propose is inspired by Wilder Penfield's account of agency-denying neural intervention:

> When I have caused a conscious patient to move his hand by applying an electrode to the motor cortex of one hemisphere I have often asked him about it. Invariably his response was: 'I didn't do that. You did.' When I caused him to vocalize, he said, 'I didn't make that sound. You pulled it out of me'.[11]

With that in mind, consider a more elaborate and clandestine intervention: A neuroscientist inserts billions of tiny networked nanocontrollers into a subject's brain, each of which functions both as a transmitter and receiver of neural signals. When the subject awakens and forms the intention to stand, the nanocontrollers intercept the outgoing signal and relay it to the neuroscientist. The neuroscientist, who wishes to test his equipment but not to disrupt the life of the subject, registers the intention and then sends a corresponding signal to the appropriate regions of the subject's brain, causing his body to move in just the ways intended. In this case we have an intention to act that is satisfied. Moreover, unlike the case of Davidson's climber, from the point of view of the subject, the intention is satisfied in just the way he would expect. Nonetheless, the ensuing movement would not be up to the subject: he moves only at the behest of the neuroscientist.

This 'nanodeviance' case presents unique problems for theories of action because it shows the futility of attempting to satisfy PC by simply adding another ingredient to the chain of causation. For any two steps in the process between the initiation and the completion of action there exists an intermediate step in the chain where a deviant cause (such as the neuroscientist's nanocontrollers) can be inserted to hijack the action. We do not need to revisit Zeno's metaphysics to make this coherent because there are spaces in the brain: the gaps between presynaptic and postsynaptic neurons. For any neural cause you choose there will be room for a sufficiently savvy neuroscientist to introduce deviance in the form of a nanocontroller that could pick up the outgoing signal and transmit it to a remote agent who, if he so chose, could transmit a signal of precisely the same type to the postsynaptic neuron.[12]

If the preceding story is coherent, it can be shown that any attempt to satisfy PC with a diachronic relationship between the motivating states of the agent and the agent's action is doomed to failure. I address several popular attempts to do just this in what follows, starting with Frankfurt's own account.

Frankfurtian guidance

Frankfurt treats guidance as a relation between agent and action that obtains for the duration of the movement; as he notes, 'action cannot be guided through the course of its occurrence at a temporal distance'.[13] This sort of guidance obtains in any purposive mechanism. Frankfurt notes that a behavior is purposive (and thus guided) when:

1   . . . its course is subject to adjustments which compensate for the effects of forces which would otherwise interfere with the course of the behavior . . .
2   . . . the occurrence of these adjustments is not explainable by what explains the state of affairs that elicits them.[14]

When these conditions are satisfied, 'the behavior is in that case under the guidance of an independent causal mechanism, whose readiness to bring about compensatory adjustments tends to ensure that the behavior is accomplished.'[15] The question then turns to whether an appeal to the nature of this guiding mechanism, with an eye toward determining whether an appeal to its participation in a theory of action can defuse deviance counterexamples.

To this end we can consider Frankfurt's example of a driver who allows his vehicle to coast downhill. In this case the coasting is under the guidance of the driver even if the driver does not actually exert any force on the vehicle. Rather, the driver guides the car in virtue of his being both positioned and prepared to intervene if necessary. This is only a loose metaphor. Frankfurt recognizes that on pain of circularity, guidance cannot itself be a kind of action, as both deliberately refraining from applying the brake and intervening to change course in light of new information are. His point is that guidance involves a process of monitoring and adjustment performed by a purposive mechanism. These mechanisms are undoubtedly necessary for action. Contemporary accounts of intelligent movement posit the existence of numerous predictive, monitoring, and adjusting mechanisms throughout the nervous system, and I will argue in later chapters that they are vital components of a full account of primitive action. However, their proper functioning does not suffice to satisfy PC.

In claiming that action is essentially agent-guided Frankfurt does not mean that there exists some additional mechanism – the agent – that exists independently of its motor controllers and that signals them into action, but his account does need an additional component to demonstrate when the purposive guidance of a mechanism is a case of agent guidance and when it is not. He is careful to note that in cases of action the *agent* does not guide the mechanisms that in turn produce action (I do not guide my motor cortex when I move), but rather guides its behavior by employing some constitutive mechanism that is itself responsible for guiding the behavior.

But this leaves the question of what distinguishes merely purposive behavior from intentional movement untouched; when do *I* count as employing one of my constitutive mechanisms? If behavior counts as being guided by the agent when some *other* causal mechanism triggers the operation of the motor controller, it seems that a source of deviance could be inserted between the information channel linking them, intercepting that signal and (if the hijacking agent so chose) relaying a signal of the same type to the activating mechanism. The case may be extended to the monitoring requirement as well: feedback from various subsystems could also be intercepted and reliably relayed.

A similar example can be motivated against George Wilson's attempt to satisfy PC by appeal to a notion of sentient direction:

> Sentient direction . . . entails that the mechanisms of the agent's bodily control, as exercised in the performance of [behavior], were systematically and selectively responsive to the agent's perception of her environment.[16]

Sentient direction, as understood by Wilson, functions in much the same way as Frankfurt's notion of guidance. Consequently, it succumbs to the same problem. The agent's bodily control mechanisms *are* exercised in the performance of action in nanodeviance cases. Consider the case where nanocontrollers are inserted at the periphery of the central nervous system, reliably and automatically relaying both information from the environment and kinesthetic feedback from movement to the relevant motor regions of the brain – the 'mechanisms of bodily control' – but intercepting and (again, at the behest of the neuroscientist) retransmitting outgoing signals. The agent's governing motor systems are properly attuned to the environment. They issue appropriate commands, receive appropriate feedback, and issue further commands as subsequent feedback dictates. Nonetheless, the agent moves because the neuroscientist commands it. The counterexample holds.

Thus it seems that the appeal to sentient direction brings us no closer to satisfying PC. This picture of guidance is that of a system that issues orders, obtains feedback, and then issues further orders as needed. This is a common way of understanding intention – John Searle notes that intentions seem to have the form of commands and that it is useful to treat them as such. But in treating the relationship between the agent's motivating states and its movement as analogous to that holding between a transmitter of orders and a receiver of orders, we find ourselves struggling with deviance worries – the gap between agent and action remains.

This objection holds against any claim to the sufficiency of any localized physical guidance mechanism for action, be it a guiding representation, an emulating system, or a monitoring and control mechanism. No appeal to prediction, signaling, or feedback will be adequate because a counterexample can always be generated where the neuroscientist intercepts and reliably transmits the relevant signal, taking the process out of the agent's control long enough to disqualify the subsequent movement as a case of action.

Readers unused to fanciful hypothetical cases may be impatient with this line of argument. But deviant causal chain counterexamples reveal that our understanding of guidance (and the related notion of control) in action is underdeveloped. It is not enough to say that behavior is guided when some state of the agent is functioning to monitor and adjust it. Nor can we simply include an ad hoc stipulation that no outside agency can insert itself into the production of our action. That sheds no new light on the question of *why* the neuroscientist's intrusion in the production of movement should disqualify it as action. I consider a few possible amendments in what follows.

Proximal intentions

One might, recognizing the strangeness of the intervention, object that the agent's guiding state (be it an intention or a subpersonal guidance mechanism) is malfunctioning somehow. The functional role of an intention, if anything, is to motivate action, not to motivate an outside agency to motivate action. Thus in each case the relevant causal mechanisms are not realizing their proper functions because some inappropriate intermediary has inserted itself between the mechanism and its behavioral output, and for this reason the movement fails to qualify as action. We might argue that a specification of proper function for a guidance system should include the clause that it be the proximate cause of its behavioral output.

Myles Brand formalizes proximate causation as follows:

e proximately causes f if

(i) e causes f and
(ii) there are no events $g_1, g_2, \ldots, g_n$ (where $n \geq 1$), distinct from e and f, such that e causes g_1 and g_1 causes g_2 and . . . and g_n causes f.[17]

Here 'e' and 'f' range over particular events. Clearly, any theory of action that treats intentions as proximate causes of movement leaves no room for deviant intervention. The specified function of a Frankfurtian guidance mechanism might also include such a clause.

This move fails for three reasons. The first, specific to the case of full-blooded intentional action, is that it is simply false that intentions are the proximate causes of movement if intentions are supposed to be distinct from the subsystems that serve to carry out their orders. Movement is a complex task, with significant work being done not only by the parts of the brain devoted to intention formation, but also by subcontrollers, which are more proximate to the movement than intentions. Thus, unless intentions supervene on both brain states and states of the musculoskeletal system, a theory of action that requires intentions to be the proximate causes of action may fail to classify any intentional movement as action.

A second objection is that we lack good reason to think that the proper functions of our guiding mechanisms ought to specify their proximity to their causal outputs. Whether the proper function of a mechanism is fixed by function over a suitable training period, function as it contributes to evolutionary success or function as dynamically specified role,[18] no such specification depends upon the idea that for a brain state to function properly it must be the most proximal cause of whatever it is supposed to effect. Even if the motor cortex could only guide body movement through some unusual intermediary process, it would surely be said to function properly if it were to do so successfully. Here the developing field of neuroprosthetics is relevant. Neuroprostheses are designed to perform motor, sensory, or cognitive functions for their users. The cochlear implant is the best-known case, but prosthetic eyes and limbs are currently being integrated into existing neural pathways. Future advances could allow neuroscientists to artificially 'patch' connections between regions of the brain, or between the central nervous system and musculoskeletal system. In such a case we would surely deny that the motor cortex was malfunctioning simply because an intermediary relay had been inserted into the process of action production.

Finally, even if the first two objections can be ignored, it seems that it a nanodeviant case could still work. As John Bishop notes in a critique of Brand's position, a controller could be added at one of the junctions between the agent's motor control centers and the receptors at her muscle fibers.[19] If an intention properly initiates a causal chain that is hijacked after the intention has done its job but before the associated motion has begun, the standard conditions for action (on the causal model) are satisfied but action has not taken place.

Alfred Mele, citing Brand, counters that it is part of the function of a proximal intention to guide and sustain action.[20] In cases of deviance, that guidance is lost, so the proximal intention view is immune to such counterexamples. But this only takes us back to where we began, for we lack an account of how this sustaining and guiding is supposed to work. Some further explication of the mechanism that bridges the gap between agent and action must be provided.

Searlean intentions in action

John Searle's account of intention in action is important to address for two reasons. Not only is it the most plausible attempt to satisfy PC of the views discussed thus far, but it also offers an attractive candidate for a theory of primitive action. Searle distinguishes prior intentions – intentions which function as the prior causes of actions – from intentions in action, which are *continuous* with actions. Searle notes, however, that although the exercise of prior intention requires the participation of intentions in action, some intentional actions do not originate in prior intentions. Spontaneous or mindless actions, for example, may be performed intentionally even if we had no reason to perform them. The distinct content of intentions in action also allows for their ascription to pre- or nonlinguistic creatures that cannot token states with the propositional content of a prior intention. In such cases, Searle argues, the relevant movement may be driven by an intention in action in the absence of a prior intention and that this suffices to secure its status as intentional action. Thus a Searlean account of primitive action would be attractive if his account of intention in action could avoid the nanodeviance objection: primitive actions could be identified with movements brought about by intentions in action without the participation of prior intentions.

Searle argues that one defining characteristic of intention is its *causal self-referentiality*: a given intention's conditions of satisfaction include the requirement that they be brought about by *that* very intention. More specifically, the representational content of a prior intention, say, to raise one's arm, can be expressed as:

> *This prior intention causes an intention in action which is a presentation of my arm going up, and which causes my arm to go up.*[21]

The presentational content[22] of the intention in action is expressed as:

> *My arm goes up as a result of this intention in action.*[23]

This additional condition sets the intention apart from other intentional states such as beliefs, which may be satisfied in myriad ways. It also precludes deviance objections. This is because Searle takes the content of an intention to be given by the conditions under which it is successfully carried out. The nanodeviance case might simply identify a further condition that must be built into the content of the intention – namely that no outside agent may intervene in the production of the relevant movement. Thus the content of the prior intention could be plausibly unpacked as follows:

> *This prior intention causes an intention in action which is a presentation of my arm going up, and which causes my arm to go up via a process that does not include the intervention of any outside agency.*

The representational content of the prior intention is not satisfied in the nanodeviance case, and thus Searle's theory can account for what has gone wrong in those cases. The success of this strategy may lead one to posit similar content for the intention in action, which is the vital component for our purposes:

> *My arm goes up as a result of this intention in action without the intervention of any outside agency.*

If this were a plausible characterization of the presentational content of Searlean intentions in action, then it would seem that his account satisfies PC. What we will see, however, is that the content of the intention in action cannot be specified in this way. Thus Searle's theory *only* skirts deviance at the level of full-blooded intentional action, and a developed account of primitive guided action must be sought elsewhere.

This can be seen by noting the difference between the *representational* content of a prior intention and the *presentational* content of intentions in action. Whereas the representational content of the prior intention is a linguistic item – a specification of its conditions of satisfaction – the presentational content of the intention in action is given by 'the experience of acting'.[24] This point is essential for deflecting an early objection to Searle's theory. Mele argues that if action depends upon the use of self-referential intentions, then young children and certain intelligent animals will be disqualified as agents because they lack the requisite conceptual repertoire.[25] Although agents may have beliefs and desires without possessing the concepts of belief and desire (a child might stand in a certain relationship to cookies without grasping the concept of that relationship), one must have the concept of an intention to have a self-referential intention: as the argument might go, to stand in an intentional relation to an object, one must have the concepts that pick that object out. Because the object of the self-referring intention includes that very intention, on the Searlean picture it seems that agents must have the concept of an intention in order to act. This sets the bar for agency unnecessarily high.

However, this objection fails to properly distinguish the contents of prior intentions from those of intentions in action. If intentions in action have no conceptual content, then their bearers need not possess any concepts to token them, and there is good reason to think that they do. Although Searle does not claim that intentions in action have nonconceptual content, Elisabeth Pacherie explicitly distinguishes concept-employing prior intentions from nonconceptual intentions in action, drawing upon fineness-of-grain arguments from the philosophy of perception.[26] This interpretation is consistent with Searle's analogy between perception and the intention in action. The intentional content of perception is provided by the visual experience, which is an unmediated presentation of the world.

Similarly, Searle notes that the intentional content of the intention in action is not a proposition, but rather the *experience* of acting.[27] This experience may include the notion that one's action is brought about by that very intention in action. This presentational content is given by the sense of agency discussed in Chapter 1: in normal cases our basic actions have a characteristic phenomenological quality. By contrast, when this process is impeded or hijacked in some *noticeable* way (as in the standard cases of deviance), we are immediately aware that something has gone wrong. This is a way in which the experience of one's acting can nonconceptually express the success and failure conditions for the production of that action. It also provides a response to Mele's objection: the actions of agents that lack the concepts of intention or action may nonetheless act when their movements are caused by nonconceptual intentions in action. '

Thus Mele's objection can be met, but only by asserting that the content of an intention in action is fundamentally different in kind from that of a prior intention. This necessitates a re-examination of the nanodeviance objection as applied to isolated intentions in action. Because our target is a notion of primitive action, we must see if the content of an *isolated* intention in action (i.e., an intention in action that does not depend on the presence of a prior intention) can be plausibly specified in a way that avoids the counterexample.

The question turns to whether the experience of acting holds constant across ordinary and nanodeviant cases of action. It seems to, for in the case of nanode-viance the agent is unaware that anything out of the ordinary has occurred. His actions are not his own, but this deceit could continue over the course of his life and he would be none the wiser. He forms intentions (in action) and acts just as intended and in such a way that the link between intention and action seems unbroken. At no point does the agent think that his intentions have been thwarted or that he is merely *trying* to move. Indeed, his intention does produce movement in the following sense: it participates in a process that culminates in the very bodily movement that it specifies, and that does so (in part) *because* it specified that movement. And in the absence of some further disqualifying detail, it seems that this sort of case meets Searle's conditions, once properly spelled out.

I have assumed that the phenomenal experience of acting picks out the content of an intention in action in such a way that the agent could register the fact of its not being satisfied. This is guaranteed by the presentational content of the intention in action. The experience of acting that Searle takes to pick out the intentional content of the intention in action is usually only disrupted in cases where the agent recognizes that something has gone wrong. Of course, there are unusual cases where the agent mistakenly believes that her intention in action has been satisfied; for example, William James describes a case of an anaesthetized patient who is ordered to raise his arm. The patient attempts to comply and believes that he has, unaware of the fact that his arm is being restrained.[28] The presentational content of the intention in action can generate false positives in cases where the relevant perceptuomotor channels have been disrupted. In cases where these channels are functioning properly, however, it seems clear that the agent will know immediately whether its intentions in action will be satisfied.[29] And if this is the case, then counterexamples that introduce in principle *undetectable* deviance into the causal chain between agent and action will not be precluded by the appeal to Searlean intentions in action.

This appeal to the conscious experience of acting may be unconvincing. Couldn't the nanodeviance case be an instance where it merely seems to the agent that the presentational content of his intention in action is satisfied, but, in fact, it is not? This would be an unusual result, because, by hypothesis, the neuroscientist's intervention would cause no neurophysiological or behavioral difference in the agent (unlike the case of James' mistaken patient). If the intentional content of an intention in action is purported to include some condition of directness but this element of the content – which is supposed to be experiential – makes no phenomenal, neurophysiological, or behavioral difference for the agent, we have little reason to suspect that any such condition is included in the content. This is true not only because it seems to make no difference in the cognitive life of the agent, but also because if the content is not presented experientially, it is unclear how it *could* be presented in a way that does not lead us back to Mele's charges of oversophistication.

It seems that our options are to (1) admit that the presentational content of the agent's intention in action is satisfied in cases afflicted with in-principle-undetectable nanodeviance, or (2) insist that some additional intentional component has not been satisfied in such cases despite being unable to provide a coherent account of it. Because the latter option seems untenable, we are left with the conclusion that nanodeviance cases hold in the case of isolated intention in action, and thus the appeal to such states does not provide the kind of deviance-free guidance we are seeking.

Pacherie's discussion of Searlean intentions in action further confirms the applicability of deviant counterexamples. She notes that the dynamical character of an intention in action must be such that its content cannot be fully determined prior to the initiation of movement. Rather, the intention in action initially expresses only a rough estimate with respect to trajectory and makes adjustments upon receiving feedback (both environmental and kinesthetic). Thus its content becomes more determinate as the action unfolds over time.[30] Pacherie refers to this feature as a form of 'dynamic indexicality'.[31]

But this suggests that the content of the intention in action can be satisfied in cases of nanodeviance. If the content of the intention in action upon the initiation of action is sufficiently indeterminate, then upon updating (with the source of deviance inserted into the process), it will have as its satisfaction condition 'This intention in action causes such-and-such an internal process that results in my arm going up', where the relevant process now includes the operations of the nano-controller. Indeed, we can imagine a more extreme case wherein *all* of the neural operations posterior to a certain point have been taken over by the alien agency. In intending to act the agent initiates an internal process with both original and replacement parts, with the alien process relaying the information that allows the intention in action to unfold over time in a way that satisfies the content of both the prior intention and the intention in action.

Guidance, deviance, and primitive agency

Although primitive action is the main focus of this book, the broad applicability of deviant causal chain counterexamples is telling. Indeed, this chapter offers a reason to think that an examination of whole-organism guidance is crucial not only to our understanding of nonhuman animal movement, but also to action theory as a whole. As Frankfurt argued, the capacity for full-blooded intentional action depends essentially on the capacity for intentional movement generally, where the core property of intentional movement is its being guided. Thus the study of deliberative action benefits from the study of this more basic capacity, which is necessary for its exercise.

What we have seen here is that this capacity cannot be satisfactorily understood in terms of diachronic relations between intention and movement. That is not to say that actions never have intentions as (contributing) causes. But the failure of traditional methods to satisfy PC does suggest that the missing ingredient – for intentional movement, and thus for its more rarified extensions – may be found outside the domain of deliberative rationality. One place to start would to be to look at simpler systems – less sophisticated, but no less interesting – to see where and how synchronic forms of whole-organism guidance emerge. We may then be able to improve our understanding of uniquely full-blooded intentional action, understood as a refinement of the 'raw material' of primitive action.

I will argue that the necessary sort of guidance cannot be traced to the operations of a single agency-conferring state within an agent, but rather must be understood as a structuring relationship between a living organism as a whole and its subordinate parts. I elucidate this claim over the next three chapters, beginning with an investigation of what motivates our attribution of behavior to whole organisms.

Notes

1 H. Frankfurt, 'The Problem of Action', *American Philosophical Quarterly* 15 (1978), pp. 157–162.
2 This claim is consistent with the idea that prior plans and intentions can set rational constraints on subsequent practical reasoning by being reintroduced as premises into a deliberative process.
3 Frankfurt, 'The Problem of Action', p. 159.
4 D. Davidson, *Essays on Actions and Events* (Oxford: Oxford University Press, 1980), on p. 79. The possibility of deviant causal chains is not in itself enough to debunk Davidson's causal theory of action specifically, since his account provides only necessary conditions for action. Thus further details may be required to bring about actions in the right way, and these details may preclude the applicability of deviant causal chain counterexamples. Presumably this is why Davidson does not find the existence of such counterexamples as devastating as Frankfurt does. But whatever these details may be, it seems clear that they must not confine the role of the motivating belief and desire to that of a temporally prior cause.
5 Even common reflexive behaviors like blinking cannot be fully explained in abstraction from the systems of which they are a part. For example, we tend to blink less often when reading something interesting. In these cases, the rate of blinking can correlate negatively with the amount of incoming information. We also blink more frequently when we are in emotional distress. Thus a full explanation of blinking must appeal to additional processes associated with, among other things, language processing and emotion. This fuzziness of the boundaries of the relevant mechanisms for blinking should not be seen as reason to reject the distinction between the behavior of agents and the behaviors of subsystems, because it seems clear that such a distinction exists no matter how complex the interactions between subsystems become.

6 Frankfurt, 'The Problem of Action', p. 157.
7 Ibid., p. 159.
8 Ibid., p. 162.
9 To use another example from Davidson: a man tries to kill someone by shooting him but misses. The missed shot startles a nearby herd of wild pigs, which tramples the target to death. In both cases the antecedent intentions are satisfied, but the progression of events is interrupted in such a way as to disqualify the rope-dropping and target-killing events as actions.
10 A. Mele, 'Intentional Action and Wayward Causal Chains: The Problem of Tertiary Deviance', *Philosophical Studies* 51:1 (1987), pp. 55–60.
11 W. Penfield, *The Mystery of the Human Mind: A Critical Study of Consciousness and the Human Brain* (Princeton, NJ: Princeton University Press, 1978), p. 76.
12 Two details must be introduced to make this story plausible. First, the neuroscientist does not suffer from any speed limitations that would introduce crippling time delays into the system (if this means that we must posit that the neuroscientist is a Martian, so be it). Second, it seems unlikely that intervention at a single synapse would matter for the production of even our most basic actions, but we might imagine millions of networked nanochips being introduced into the brain of a hapless subject such that they could work together to affect significant changes in the subject's behavior if his controller so chose.
13 Frankfurt, 'The Problem of Action' p. 158.
14 Ibid., p. 160.
15 Ibid.
16 G. Wilson, 'Reasons as Causes for Action', in G. Holmstrom-Hintikka and R. Tuomela (Eds.), *Contemporary Action Theory* (Dordrecht, NL: Kluwer), pp. 65–82, on p. 146.
17 M. Brand, 'Proximate Causation of Action', *Philosophical Perspectives* 3 (1970), pp. 423–442, on p. 425.
18 These views are attributable to Dretske (1981), Millikan (1993), and Bickhard (2004), respectively.
19 J. Bishop, *Natural Agency* (Cambridge, UK: Cambridge University Press, 1989).
20 A. Mele, 'Recent Work on Intentional Action', *American Philosophical Quarterly* 29 (1992), pp. 199–217.
21 J. Searle, *Intentionality: An Essay in the Philosophy of Mind* (New York: Cambridge University Press), p. 95.
22 Searle treats the relationship between prior intentions and intentions in action as analogous to the relationship between memories and visual experience: prior intentions represent their contents, whereas intentions in action directly present their contents. Analogously, visual experience is a direct presentation of the world to the perceiver, whereas memory is a re-presentation of that world. The directness of this relationship between the intention in action and the corresponding movement is what makes the intention in action and movement a candidate for closing the gap between agent and action.
23 Ibid., p. 93.
24 Ibid., p. 88. Here Searle is referring to the content of intentions in action alone, as evidenced by the fact that in the passage quoted he draws an analogy between this content and what he calls the presentational content of visual experience.
25 Mele, 'Intentional Action and Wayward Causal Chains: The Problem of Tertiary Deviance', pp. 55–60.
26 E. Pacherie, 'The Content of Intentions', *Mind & Language* 15:4 (2000), pp. 400–432. Such arguments claim that the content of visual experience is more detailed than any description of that experience can capture. Similarly, the intention in action provides continuous, dynamic, fine-grained guidance over time. Its content picks out a process that is too detailed to be captured conceptually and need not be in order to function.

27 Searle notes that the experience of acting need not be conscious or phenomenal. Indeed, he notes that the intention in action and the experience of acting share intentional content; they differ only in their phenomenal properties, which are inessential to that content.

28 W. James, *The Principles of Psychology, Volumes 1 and 2* (Boston: Harvard University Press, 1890), note p. 489.

29 Any appeal to 'proper function' is contentious, and indeed I briefly argue against appealing to proper functions in a defense of 'most proximal cause' theories of intention. I will address this in more detail in a later chapter, but for now it seems reasonable to say that a subject who is unable to fully integrate perceptual and kinesthetic information in the service of acting is not functioning properly as an agent.

30 Searle is explicit about this point only with respect to prior intentions as incomplete plans, but it seems clear that something similar must be the case for intentions in action to perform the task he has set for them.

31 'Indexicality because the exact value of certain constituents of the representation (amount of force that needs to be programmed, precise shape of the hand, and so on) can only be fixed relative to a context. Dynamical indexicality because the context itself must be brought into existence by the intention in action.' Pacherie, 'The Content of Intentions', p. 414.

3   Whole-organism agency

Of parts and wholes

Thus far I have argued that action is a broad behavioral kind common to humans and nonhuman animals alike, of which human deliberative action is a complex subkind. The core features of this broader sort of 'primitive action' are that agents (rather than mere agent-parts) produce it and that it is a form of coordination or guidance. Frankfurt's account of intentional movement as whole-organism guidance was an ideal starting point for fleshing out the details of primitive action, but guidance as conceived by Frankfurt and others is an inadequate foundation for a theory of action. Specifically, the necessary sort of guidance cannot be reduced to the operations of a monitoring/adjustment mechanism. Such mechanisms undoubtedly *are* essential to the performance of action in many agents, but their operation is insufficient for guidance of the agency-conferring sort.

The problem is that we have attempted to answer the question of what whole-organism guidance could be without first understanding what it could mean for a *whole organism* to do anything at all. In discussing mechanisms of monitoring and control, we are restricting our focus to the operations of *parts* of organisms, with the hope that perhaps certain parts could be identified as more properly 'of the organism' than others. But this is a mistake. It is only within the context of a whole organism that the function of any particular behavior-producing part can count as a constituent process of whole-organism guidance. Indeed, I will argue in later chapters that it matters a great deal that the relevant mechanisms function within the context of a whole, *living* organism. To see this, I turn to Frankfurt's second challenge for the philosophy of action: to establish the conditions for attributing action to an agent rather than to one or more of its submechanisms. To this end I will consider the merits of several views of agency and behavior, each of which offers a partial solution to the problem. We will see that an adequate answer to this question will require a perspectival shift from mechanism to structuralism (or, if one prefers, from reductionism to holism).

The distinction between organism-guided and part-guided movement is a variant of the commonsense distinction between things the agent does and things that happen to it; it is the distinction between things the agent does and things that happen *within* it. It partially informs our distinction between activity and passivity.

I am passive with respect to my hiccups, for example, but active with respect to my feigned hiccups. The processes of running and having a seizure are both internally generated, but my dog actively runs, whereas he is the passive victim of his epilepsy. Thus we need to understand what conditions make it appropriate for us to say that my dog is running rather than being passively propelled through space by his constituent parts. In keeping with developing a theory of primitive agency, it will also be valuable to determine at what level of complexity such attributions are anything more than metaphor. Can ants perform whole-organism movements? Can amoebae? Or do we merely extend the language of agency to them for the sake of convenience, or because of some superficial similarity between their behavior and the behavior of *genuine* agents such as ourselves?

Tyler Burge jogs our intuitions regarding what sorts of movements are whole-organism and which are not:

> A spider pursues, jumps on, bites, and eats its prey, approaches and insemi-nates its mate, navigates past an obstacle, or runs across a web. These actions are distinguished from processes occurring only within the spider. The spider ingests; its stomach digests. Only sub-systems operate in the circulation of fluids, and production of protein, semen, or wastes . . . An amoeba's ingest-ing its food is action. Digesting its food is not. A paramecium's swimming forward or backward is action. The plasmolysis that causes shrinking of the paramecium in highly concentrated solutions is not. The crawling of a tick towards a heat source is active and attributable to the whole organism. Pro-tein transfer through its membranes is not.[1]

The passage's intuitive plausibility may be due to the fact that an active/passive distinction is implied in our linguistic practice: *I* eat, whereas *my stomach* digests. Of course, the appeal to language only takes us so far. We also say that Max collapses, Kasey sneezes, Karolina blushes, and Drew sweats, but these are not cases of action – the descriptions are shorthand for processes that happen to agents by way of issuing from processes within them. Certain borderline cases are especially vexing: Jessica sleepwalks; Shane flinches despite himself; Mike controls *where* he defecates but not *whether* he defecates, and so on.

Indeed, our ascription of actions to whole organisms may be guided by mere convention rather than any deeply held folk intuitions about agency. For example, Burge notes that we tend to attribute certain peripheral behaviors to agents as wholes (Cody blinks; Sarah coughs) if they engage parts of the body that feature prominently in our interpersonal lives (in this case, heads and faces).[2] But we recognize that sneezing and coughing are not usually actions. In short, a term can occupy the subject position of an action sentence without its referent actually being the agent of the event. Language can orient us, but the real work must be done at the level of physical processes.

What we find at this level, however, is that the agent seems to vanish altogether. Organisms are composed of numerous guiding mechanisms, none of which can be identified with the agent. Nor is there any particular mechanism that, when active

in the production of movement, obviously establishes it as action. We might identify regions of the brain responsible for storing and running motor schemata, of
course, but the deviance examples and the case of the alien hand syndrome sufferers demonstrated their activation is insufficient for action. Despite our advances
in the neuroscience of behavior production, the agency-conferring region of the
brain eludes us. And this is to say nothing of organisms that lack brains altogether!

To complicate matters further, only some of a given agent's parts contribute to
the production of any of its actions. It would be fallacious to require that *all* of
an agent's parts be active in the production of a movement for that movement to
count as active. Other parts make no contribution, either because they have not
been activated or because their potential contributions to the action have been
suppressed. Nonetheless, when I raise my arm, the arm-raising event is an action
properly attributed to me rather than to my arm or the combination of my arm and
the relevant parts of my brain and central nervous system. But it seems that only
the exercise of certain capacities will bring about processes worth describing as
whole-organism behavior: the chain of processes culminating in arm raising must
be distinct in kind from that involved in sweating or hiccupping.

One might worry that the distinction is largely arbitrary and that the vagueness
of our language about whole-organism behavior results from there being no real
fact of the matter as to whether an organism as such is acting. If there is not, then
it seems that the matter will be settled by explanatory convenience. A study of
animal sociality, for example, may address the operations of a system that takes
whole organisms as its primitive elements. A finer grain of analysis would unnecessarily complicate such a study. Questions as to whether the whole organism is
really acting get no traction on such a picture.

There are three reasons to reject this subjectivist way of thinking. The first is
that the structure of the universe does not answer to our explanatory and predictive needs. A marketing analyst may find it useful to treat the set of 18- to 34-year-
old males as if it were a single entity that *demands* or *rejects* various products, but
we do not thereby presume that they constitute such an entity. The current philosophical debate regarding the collective agency is another case. Perhaps groups
such as corporations and voting constituencies can be agents, but the matter will
be settled by their organizational properties, not by whether it is expedient for
outsiders to describe them as if they were.[3]

The second reason to suppress any antirealist impulses is that the notion of
the whole organism has taken on renewed explanatory significance in biology.
Reductionist/mechanistic approaches to biology held sway throughout the twenti-
eth century, bolstered by the success of population genetics and molecular biology.
The explanatory success and unifying power of these fields led many biologists
to adopt reductionism as a guiding methodological principle. Richard Dawkins
expressed this view artfully by describing the living organism as little more than
a 'survival machine', created by its genes for the sole purpose of fulfilling *their*
aim of self-transmission.[4] On such a view the organism itself explains nothing,
serving as a mere 'interface between the phenotypic expression of genes and the
selecting role of the environment'.[5]

Recent developments in biology have suggested that the genocentric approach has its limits. Indeed, two of the most significant results of the Human Genome Project were that (1) humans have roughly one-fifth the number of genes anticipated (a relatively scant 19,000), meaning that organismic complexity is not a function of gene number, and (2) we cannot predict an organism's phenotype from even a comprehensive understanding of its genotype. This is due in part to the fact that the information flow from genotype to environment is not one way – many organisms *construct* their environmental niches, thereby changing the very environmental pressures that affect gene expression and selection.[6] It is also due to the fact that DNA transcription takes place in the context of many other processes, including RNA transcription, protein–protein and protein–DNA interactions, and metabolic processes, all of which may have effects on gene expression. This has prompted the rise of holistic 'systems' approaches to biology, which stress not only the interdependence of these intraorganismic systems, but also the explanatory importance of the organism itself, which has been reprioritized not as the causal output of its genes, but rather as the structuring organization within which (and, as we will see, *for which*) these processes operate. No longer a mere epiphenomenon of its parts, the whole organism is both *explanandum* and *explanans* from the perspective of contemporary systems biology.

This leads us to the third reason to take whole-organism talk seriously. If whole-organism behavior could be shown to figure centrally in both ethology and action theory, connections may be drawn between the biological, psychological, and moral domains of inquiry. The philosophical value of such a union will be apparent to those of us who agreed with Wilfrid Sellars' claim that the aim of philosophy is 'to understand how things in the broadest possible sense of the term hang together in the broadest possible sense of the term.'[7] The unifying potential of an account of whole-organism behavior makes the attempt to produce such an account worthwhile.

Burge on primitive agency

Tyler Burge's account of primitive agency is written in the spirit of Frankfurt's work and thus serves as an ideal starting point for the present investigation. Burge's investigation is an attempt to make sense of the phylogenetic background conditions constraining animal perception. One core thesis of this work is that preperceptual forms of agency partially determine the class of objects that are candidates for perception in more sophisticated agents. A second is that the norms governing perception and representation in agents like us are closely related to more basic 'natural norms' governing processes in very basic forms of life. Burge's analysis is thus a helpful starting point for understanding basic forms of agency.[8]

Burge offers the following characterization of primitive agency:

> The relevant notion of action is grounded in functioning, coordinated behavior by the whole organism, issuing from the individuals' central behavioral capacities, not purely from sub-systems.[9]

Burge identifies preperceptual taxis as the earliest form of primitive action, contrasting it with nonactive kinesis and tropism. Taxes, understood as 'active orientations' of 'freely motile organisms', are nonrandom forms of locomotion that orient agents in ways that contribute to their flourishing and survival. In so contributing, taxes are forms of purposeful motion. They are common in single-cell eukaryotes that are capable of steering themselves toward and away from various stimuli. As Burge puts it, these motions 'depend on the condition of the organism and are produced by the release of forces characteristic of the organism'.[10] By contrast, the random tumbling of bacteria, triggered by changes in chemical gradients in its environment, is not directed or coordinated by the bacterium. Tumbling is something that *happens to* the bacterium, not something it *does*. This is not to say that the tumbling is not valuable – it helps to keep the bacteria in favorable chemical gradients and thereby contributes to the success of its active taxis – but its contribution is that of a passive process, however necessary.

Burge's account of primitive agency is a useful starting point, but Burge himself never intended to offer a definition of the kind, using it instead as a launching point for his study of perception. Indeed, his examination bottoms out in what is at best a pragmatic appeal to behavior as a primitive unit of explanation in ethology, and at worst the view that our attributions of agency to more basic systems are mere metaphorical extensions on the basis of perceived behavioral similarities to less controversial cases:

> I doubt that there is an independent criterion for whole-individual agency. Again, the fact that the amoeba is eating seems to carry as much weight in the judgment that the eating is active rather than passive behavior as the fact that there is coordination with the individual's central capabilities . . . I think that our understanding of these matters is probably partly guided by an antecedent list of whole-individual functions that already embody conceptions of activity by the whole individual organism – eating, navigating, mating, and so on.[11]

Nonetheless, Burge's account provides us with three inroads toward understanding primitive action: (1) it is whole-organism functional behavior; (2) it issues from the central behavioral capacities of a system; and (3) it is the result of a release of forces characteristic of the organism. This characterization is lacking but can be amended in useful ways, and its examination will provide a way forward for a more developed account.

Does whole-organism behavior require centralized processing?

What does it mean for a behavior to issue from the central behavioral capacities of an individual? What are these capacities, and when can we correctly say that they are being exercised? Burge offers several criteria, the first being that such behavior is endogenously driven. Agents are not simply pushed around by their environments: they make energetic contributions to their actions. By contrast, reflexive stress, or schreck, responses passively shut down central behavioral capacities and thus are not instances of action. Of course, much hangs on the nature of the energetic contribution, as we have seen that the mere fact that an internal process produces an agent's movement is hardly enough to establish its status as active movement of any kind.

One way to identify when a movement is attributable to an organism as a whole rather than to one of its parts might be to identify a central controller. A hard distinction could then be made between central and peripheral systems within the organism, with the centralized systems prioritized as 'organism-level' and attributing the movements they produce to the organism as a whole. Burge does much to motivate the idea that whole-organism behavior requires control on the part of the central nervous system (CNS):

> [I]n the cases of larger animals, there is usually a fairly clear distinction between central and peripheral processes that correlates roughly with an anatomical distinction between processes that are controlled by the central nervous system and those that are not.[12]

The classical reflex arc, for example, does not pass through the CNS and is thus deemed 'peripheral'. But Burge also recognizes that the presence of a nervous system as such is not necessary for action – the amoeba engulfs its food but it is not directed by anything like a brain. Rather, he argues that the difference hangs on whether something in the organism performs the essential role of the CNS, namely that of *coordination* across subsystems, and more specifically, sensorimotor subsystems. I think sensorimotor coordination is a vital piece of the puzzle we are attempting to solve, but we will see in what follows that the nature of coordination is no less murky than the notion of guidance explored earlier.

The appeal to coordination is attractive – if behavior is the product of multiple subsystems operating in tandem, it cannot be attributed to the operations of any one of them – but we need look no further than the immune system to see that coordinated intersystem processes frequently operate at suborganismic levels.[13] Restricting our focus to sensorimotor systems allows us to limit our investigation to the mechanisms most directly responsible for movement, but when is it appropriate to say that two or more processes – sensorimotor or otherwise – are coordinated, rather than merely operating in tandem? The appeal to functionality cannot itself settle the matter, for it might be quite good for an organism to instantiate a series of parallel processes that, although operating in ways that satisfy the relevant functional norms, are not coordinated with each other.

In the face of this worry we might think that we must explain coordinated sensorimotor activity by appealing to some mechanism that has the role of coordinating subordinate systems, such as the central nervous system. If this is the case, then perhaps Burge's 'central behavioral capacities' truly do depend upon the participation of a central control mechanism of some sort.

Coordination without coordinators

The question turns to this: Does whole-organism coordinated behavior require the contribution of a control mechanism? A number of famous examples have given cognitive scientists good reason to think that it does not. Rodney Brooks' robot Herbert retrieved empty soda cans in an unpredictable real-world environment, but did so without the use of a central controller to manage inputs and guide behavior. Herbert embodied a layered subsumption architecture, each layer of which performed a simple task (e.g., wandering, avoidance, grasping) when it was active. Each layer would override the others when certain features of the environment activated it. Thus, although Herbert lacks an internal controller – Brooks describes it as 'a collection of competing behaviors'[14] – it is capable of coordinated activity.

Developmental studies of movement also undermine the claim that coordination requires a central controller by showing our tendency to posit central control points where none actually exist. Until fairly recently, analyses of rhythmic infant kicking posited the existence of a central pattern generator (CPG) that might, upon maturation, serve as a controller for adult locomotion. However, Esther Thelen and Donna Fisher's landmark study demonstrated that no such CPG was operative. If a CPG had been governing the rhythmic kicking of the infant, an electromyography (EMG) study should show alternate activation of the flexor and extensor motor neurons in the legs during kicking. But EMG results showed that the infants' flexors and extensors contracted simultaneously at the initiation of flexion and did not contract at all during the extension portion of the kick.[15] The form of the kicking pattern was not determined by the motor control system, but rather emerged through the joint contributions of the infant's nervous system, the natural springiness of the infant's neuromuscular apparatus, and the constraints introduced by gravity.[16]

This suggests that coordinated behavior need not depend upon the control of a coordinating system, but can dynamically self-organize from multiple interdependent factors involving nervous systems, bodies, and environments. Other studies have shown that coordinated patterns of behavior can be generated in infants by adjusting the values of various parameters in the nervous system–body–environment system in which it emerges: for example, infants could produce mature stepping patterns when suspended on moving treadmills[17] or submerged up to the neck in warm water.[18] In each of these cases behavior emerges from the co-contribution of nervous system, the physical properties of the body, and those of the surrounding environment.

Work on decerebrate cats has also demonstrated the distributed nature of movement production.[19] These cats can achieve natural walking and running gaits when suspended over a moving treadmill. Although they cannot initiate their movement, they naturally transition from pacing gaits to diagonal "trotting" and back as a function of treadmill speed. The cat's body obeys a general rule: if an unloaded leg is extended, it will swing. It is able to do this in the absence of executive control because a suborganismic mechanism – the medial medullary reticular (MMR) formation – remains intact. The MMR integrates proprioceptive feedback from the Golgi tendon organs (unloading) and velocity-sensitive flexor muscles (extension) to produce rhythmic movement. This system 'selects' from energetically appropriate gaits on the basis of this biomechanical feedback.[20] This work suggests that behaviors such as running *do* require integration of sensorimotor data, they do *not* require the degree of executive control one might typically assume, and that large-scale shifts in coordinated activity such as gait transition can require none at all.

The octopus has recently received attention as an intriguing example of a massively distributed system. The octopus' body structure presents unique control problems – it has eight long arms that it can stiffen, elongate, and bend in any direction and at any point along their length. This amounts to a nearly infinite number of degrees of freedom (contrast this with the human leg, the bone and joint structure of which mercifully restricts its movement options) and a seemingly intractable problem for the octopus motor system.

Although work on octopus interlimb coordination is still in its early stages, recent work has shown that the octopus evolved to solve this problem by distributing its workload – over half of its 500 million nerve cells are found in the peripheral nervous system of its arms. Each arm is in fact a semiautonomous sensorimotor network, and many single-arm activities (such as grasping, which requires division of the limb into proximal, medial, and distal segments from the point of contact) are fully coordinated within the neuromuscular arm networks.[21] The higher motor centers are dedicated to coordinating complex interlimb movements, but even here their coordination does not amount to full control; local stimulation to octopus motor centers always activates multiple arms at once, suggesting that they lack somatotopic organization and can only run multilimb motor schemata.[22] Indeed, in the case of coordinated octopus behavior such as crawling, the central nervous system determines *which* of the arms will be active at a given time but does not prescribe *how* they will behave.[23] It does this by integrating information from the peripheral, visual, and vestibular systems, but not by controlling or prescribing its behavior in the central way that a driver might control the movement of a car.

There is also good reason to think that patterns of coordinated activity may not be fully prescribed by *any* mechanism within the agent. This has been most carefully explored in the design and study of artificial biological agents, or 'animats'. For example, Auke Jan Ijspeert evolved a simulated salamander that transitions from walking to swimming patterns as a function of tonic environmental input, and does so without changing the organization or activity of the relevant behavior-producing circuitry.[24] The salamander's internal pattern-generating mechanism prescribes neither walking nor swimming, but rather generates activity that, in water, amounts to swimming and that, in land, amounts to walking. The result of this highly efficient system is that the environment 'selects' adaptive behavior from the salamander. The actual behavior cannot be explained by the salamander's internal workings alone, but requires the inclusion of interactions between its nervous system, its body, and its environment.[25]

These cases suggest that in many systems the workload of action production is distributed rather than centralized, and that although the operations of the central nervous system are certainly necessary for action in many organisms, its role is that of a co-contributor to a complex process of interaction between nervous systems, bodies, and environments. In such systems it is increasingly difficult to identify a single mechanism that could be said to function as *the* central controller of the movement.

The possibility that centralized control is unnecessary for agency has motivated some philosophers to argue that it may not obtain in human beings. Daniel Dennett argues that our apparent unity of consciousness is illusory and that we may in fact be massively decentralized, self-organizing systems akin to termite mounds, slime molds, and anthills.[26] Andy Clark contends that studies like those mentioned earlier suggest that we ought to reject the traditional picture of action as issuing from a central controller in favor of an ecological picture of action whereby the 'self' (or selves – Clark is not committed to there being a single hub) occasionally intervenes to 'nudge' our dynamics into or out of attractor fields in a mathematically specifiable phase space. But this is not enough to designate the 'soft self' as the central controller of action, nor does it distinguish the activities of the whole organism from the activities of its parts.

All of this is to say that whole-organism behavior cannot be explained by appealing to the operations of a central control mechanism. It seems unlikely that *all* of the relevant guidance associated with the production of a given movement can be traced back to a single mechanism in such a way that we might identify a central homunculus as the true agent (or as the agency-establishing mechanism). None of this gives us the ability to make sense of the original question: When is the agent acting rather than some of its parts? Clark considers what we are left with on this picture:

> There is *no self*, if by self we mean some central cognitive essence that makes me who and what I am. In its place there is just the 'soft-self': a rough-and-tumble control-sharing coalition of processes – some neural, some bodily, some technological – and an ongoing drive to tell a story, to paint a picture in which 'I' am the central player.[27]

But this leads us immediately back to the problem under consideration: if this picture of even a sophisticated system such as the human person is correct and sensorimotor coordination is not handled centrally, then when, if ever, should we think that actions are attributable to a single, unified agent rather than to some of its parts?

Is functional behavior necessarily whole-organism?

Here an examination of the functional character of *behavior* may bring us closer to our goal. Burge argues that functional behavior (1) carries with it the notion of a norm of proper function and (2) is 'whole-organism' in that its function is attributed to the individual, not to any of its subsystems. He appeals to an evolutionary notion of biological function[28] to differentiate whole-organism function from sub-system function: whole-organism functions are frequently explained in terms of their specific contributions to organismic fitness or fecundity. Burge also notes that some whole-organism functions may be explained by appealing to the broader notion of a 'naturally flourishing life'. Thus some whole-organism functions may not have direct evolutionary payoffs but may nonetheless contribute to 'realization of the animal's life course and natural biological capacities'.[29]

None of this isolates the sort of whole-organism function that is relevant for primitive agency, however. The fact that a mechanism performs a function, the description of which references the whole organism, is not enough to establish its behavioral output as the organism's action. My heart beats because I need it to (its general functional description references the whole organism), but I do not beat my heart (the whole organism is not performing the function). Thus we cannot appeal merely to functional descriptions to determine when we have found a whole-organism function of the agency-establishing sort. We must first examine cases where a movement can be attributed to the whole organism and then determine in what sense that movement is functional. In other words, we must distinguish cases when movement is functional *for the organism* from cases when *it is the organism that is functioning.* Only the latter are relevant for purposes of understanding agency.

The appeal to behavior is a plausible first step for distinguishing the proper sort of whole-organism function. Many of the event types commonly associated with behavior – jumping, eating, running, grooming, hiding, and so on – are things *animals* do. When a spider pursues its prey, the pursuit is the spider's behavior, not its legs' behavior.[30] It seems intuitively awkward to refer to a part of the spider as *behaving* at all. Thus an examination of the conditions for behavior might provide a first step toward understanding the conditions for primitive action. So what makes movement count as behavior? Here I assess accounts of behavior from Ruth Millikan and Fred Dretske. We will see that although neither account is adequate for making sense of whole-organism function, both have valuable components that will play a role in a positive account of primitive action.

Ruth Millikan distinguishes behavior 'from mere motions, from incidental effluences of the organism, and from other incidental changes occurring on its surface'[31] by distinguishing functional behavior from mere state change. She argues that an animal's bodily movement may fall under an indefinite number of descriptions, but that the ethologist should only be concerned with those descriptions that make reference to 'behavioral laws' that appeal to the functional character of the movement. Millikan lists three necessary conditions for behavior:

1   It is an external change or activity exhibited by an organism or external part of an organism.
2   It has a function in the biological sense.
3   This function is or would normally be fulfilled via mediation of the environment or via resulting alterations in the organism's relation to the environment.[32]

The disjunctive first condition suggests that Millikan's concept of behavior is too broad to adequately characterize *whole-organism* behavior, but her account has several interesting features that are worth considering as necessary conditions. The first is that behavior is necessarily functional. To use Millikan's example, the mouse runs away from the cat, but it can also be described as running toward an incoming broom, closer to London, away from magnetic north, and so on. The description that matters for the ethologist is the first: the movement is best characterized as evasion because it is good from an evolutionary perspective for the organism to evade predators. This seems to provide a principled reason for treating certain behaviors as whole-organism: if the organism is the relevant unit of selection in evolutionary explanation, then it seems right to say that (with some qualification) behaviors that contribute to evolutionary success may be properly viewed as whole-organism.

Despite these virtues, the appeal to evolutionary teleofunction fails. Evolutionary explanations are biologically useful, but ultimately an explanation of a system's behavior must be grounded in the operations of the system, not in the processes that brought that system into being. Mark Bickhard offers the case of two lions, the first produced through generations of evolution and the second spontaneously generated as a molecule-for-molecule replica of the first (perhaps in swampman-esque fashion).[33] The two lions would be qualitatively identical and thus would behave in similarly identical fashion, but only the first lion would have historically specifiable functional states. Bickhard takes this result to show that functional states, so assigned, are causally impotent and thus can play no role in explanations of the actual behaviors of organisms.

A defender of the historical approach might reply that the spontaneously generated lion *does* in fact have functional states in virtue of its being identical to an organism that *was* produced through evolutionary selection pressures. But this response is unsatisfying for two reasons. First, it leads us into the murky territory of treating the relationship between functional states and evolutionary history as something quite vague – the spontaneously generated lion was not produced by evolution, but its physical states have the functional states they do because they *would have been* produced in that way. Second, it does not make sense of the alternative case of an entirely novel creature (call it Swamp-anomaly) that does not share the physical architecture of any existing object. Swamp-anomaly is physically possible, but would neither have an evolutionary history of its own nor poach one from some structurally isomorphic creature. But if this novel agent were to, for example, perceive an object and run away from it, we would surely wish to ascribe the corresponding functional behavior to it.

This is not to say that evolutionary explanations are irrelevant to the life sciences. It is only to say that grounding the proper function of a system in evolutionary history is, at best, an indirect means of characterizing it. As Fred Dretske notes:

> [E]volutionary (or phylogenetic) explanations of behavior, whether the behavior of plants or animals, are best understood, not as supplying structural causes for the behavior of today's plants and animals, but as causal

explanations for why there are, today, plants and animals that are structured to behave this way.[34]

Although Millikan emphasizes biological function, her third criterion is crucial for the present purposes. Millikan argues that certain broadly functional state changes – callus formation, tanning, shivering – do not count as behavior because they 'do not effect changes in, or in relation to, the environment in order that the environment should give a return on the investment'.[35] The metaphor of a return on investment is helpful: the organism is viewed as expending energy in the service of acquiring something from its environment. Thus Millikan rules out the case of the clam that slows its activity in response to decreases in ambient water temperature on the grounds that 'this will not be a strategic deceleration but a mere byproduct of the organism's chemistry'. Movements must occur 'more often than randomly . . . when the environment is ready to cooperate'. Thus behavior must be specified relative to a particular environment. I will revisit this idea of investment in Chapter 6.

Millikan's overall view of behavior can be summarized as follows: behavior is specified by appealing to functional laws, and those laws must make reference both to the organism (because the relevant kind of function is performed in the service of the organism – for Millikan, its fecundity) and to its environment (because the behavior must effect changes in the environment that in some way provide a return on energetic investment). She concludes that the study of behavior is necessarily the study of organism–environment systems; to understand the organism is to understand it as *embedded* in a particular environmental niche.

All of this seems useful for characterizing primitive action. Unfortunately, it is inadequate: Millikan attributes behavior to organisms both when the organisms generate movement and when their parts generate it (her first criterion). Thus her account cannot help us to distinguish whole-organism functional behavior from functional behavior that occurs within an organism. But it does seem plausible that primitive action is a subkind of behavior and thus would share many of its features.

Dretske begins *Explaining Behavior* with the now-familiar distinction between the things a system does and the things that happen to it, ultimately defining behavior broadly as a process that originates from within the organism in some special way. The way in which the process originates is important to distinguish the sort of behavior we attribute to animals from the sort of behavior physicists might attribute to electrons; there is a sense in which the way an electron swerves in a magnetic field qualifies it as something the electron does in virtue of the kind of object it is. Dretske takes this fact to motivate a demand for a minimal level of internal complexity in the system 'so as to make the internal-external difference reasonably clear and well-motivated'.[36] As Dretske notes, there is no clear difference between what an electron does in a magnetic field and what happens to it in an electron field. By contrast, there is a difference between what happens to an animal placed in water and what it does when so placed.

As with Millikan's account, Dretske's characterization of behavior is useful but is too broad to capture what is unique about *whole-organism* behavior. As Dretske notes, tree leaf shedding meets his criteria. Phototropism in plants also makes the cut, as do the behaviors of vacuum cleaners. Even the basic knee-jerk reflex may count, as it involves an asymmetry in energetic contribution to the movement: 'the stimulus acts like a "releasing force" on the organism in the sense that the energy expended in the response far exceeds the energy provided by the eliciting stimulus'.[37] This means that passive klinokinetic tumbling in bacteria would also count as behavior, as would a seizure triggered by a flashing light.

Indeed, Dretske's view is inconsistent with the idea that a behavior could be objectively caused by *the organism as a whole* as opposed to one of its parts because he denies that there are objective facts about the primary causes of behavior. This is because every instance of behavior has innumerable contributing causes, none of which can be identified as *the* cause of the behavior. Dretske uses the example of *lordosis*, the back arching performed by female cats in the presence of males. Lordosis is coordinated body movement, endogenously driven, and is something we would ordinarily attribute to the cat, not its parts. But the cat assumes this posture only under certain conditions: when it sights a male, when it is in heat, when it is not underwater, etc. Dretske suggests that the vast number of contributing causes of a given behavior indicate that no one cause could be pinpointed:

> Unless there is a principled way of saying which causal factor is to be taken as *the* cause of movement or orientation, the present system of classification provides no principled way of saying whether the cat is *doing* anything. It gives us no way of telling what is, and what isn't, behavior.[38]

Dretske concludes that the matter of whether a given movement counts as behavior ought to be settled by the explanatory interests of the scientists studying it. Indeed, *Explaining Behavior* is intended only to be an analysis of those processes that ethologists and psychologists already classify as behavior. Dretske is not interested in what *must* be classified as behavior, but rather what *can* be. And the range of endogenously driven processes that *can* be classified as behavior include bodily movements produced by peripheral subsystems of the organism. This makes sense of Burge's suggestion that the allegedly basic fact that the amoeba is eating ought to take precedence over whether any particular set of criteria is satisfied. If behavior is taken as an explanatory primitive in ethological explanation, then the interests of ethologists will determine which behaviors are whole-organism and which are not. But if we are seeking an account of whole-organism behavior that does not depend upon our interpretive practice, we will not find it here.

This reflection on behavior has revealed a number of criteria for whole-organism behavior (and thus, subcriteria for primitive action). In particular, whole-organism behavior is:

1 Functional behavior in some sense of 'functional'.
2 An essentially embedded process involving energy exchange between the organism and its environment.
3 Endogenously driven movement.

None of this is sufficient for whole-organism behavior, however, and one worries at this point that we have turned over all of the available stones in our search for a complete account. If we have, then perhaps there *is* no defining characteristic of whole-organism behavior and we ought to adopt an antirealist position about behavior, and consequently about primitive action.

Thus far I have attempted to flesh out the core elements of Burge's account of primitive action. This attempt has generated some useful criteria for action but has not generated any sufficient conditions for action. This may, of course, be because there aren't any, in which case we ought to talk about the features of our interpretive practice. But the limited success of this analysis may be due to the fact that I have not yet addressed the problem in the right way. Throughout this investigation I have appealed to mechanistic explanations (seeking a guidance mechanism, attempting to make sense of the organism as the efficient cause of its behavior) and historical explanations (teleofunctions grounded in the evolutionary history of the organism). Neither has been satisfactory. But a third way is available – a structural, holistic perspective that takes seriously the role of the organism as a whole.

Notes

1　T. Burge, 'Primitive Agency and Natural Norms', *Philosophy and Phenomenological Research* 79:2 (2009), pp. 251–278, on p. 256.
2　Ibid., p. 262, note 17.
3　For one interesting attempt at just such an account, see C. List and P. Pettit, *Group Agency: The Possibility, Design and Status of Corporate Agents* (Oxford: Oxford University Press, 2013).
4　R. Dawkins, *The Selfish Gene* (Oxford: Oxford University Press, 1976), p. xxi.
5　D. Nicholson, 'The Return of the Organism as a Fundamental Explanatory Concept in Biology', *Philosophy Compass* 9:5 (2014), pp. 347–359, on p. 348.
6　J. Odling-Smee et al., *Niche Construction: The Neglected Process in Evolution* (Princeton: Princeton University Press, 2003).
7　Wilfrid Sellars, 'Philosophy and the Scientific Image of Man', in R. Colodny (Ed.), *Frontiers of Science and Philosophy* (Pittsburgh, PA: University of Pittsburgh Press, 1962), p. 35.
8　Burge denies that his account is a definition. The remainder of this chapter should not be viewed as a refutation of his view, but rather as an elaboration of it, although perhaps not one that Burge himself would endorse.
9　Burge, 'Primitive Agency and Natural Norms', p. 260.
10　Ibid., p. 259.
11　Ibid., p. 264.
12　Ibid., p. 263.
13　It has, however, been suggested that the human immune system may partially constitute a unique sort of selfhood. Insofar as the immune system constitutes a kind of self, its operations *may* be viewed as things it does. This is irrelevant to the question of what constitutes organism-level action, however.
14　R.A. Brooks, 'Intelligence without Representation', *Artificial Life* 47 (1991), pp. 139–159, on p. 149.
15　E. Thelen and D.M. Fisher, 'The Organization of Spontaneous Leg Movements in Newborn Infants', *Journal of Motor Behavior* 15 (1983), pp. 353–377.
16　E. Thelen and L. Smith, *A Dynamic Systems Approach to the Development of Cognition and Action* (Cambridge, MA: MIT Press, 1994), p. 79.

17 E. Thelen and B.D. Ulrich, 'Hidden Skills: A Dynamic Systems Analysis of Treadmill Stepping During the First Year', *Monographs of the Society for Research in Child Development, Serial No. 223* 56:1 (1991).
18 E. Thelen, R. Ridley-Johnson, and D.M. Fischer, 'Shifting Patterns of Bilateral Coordination and Lateral Dominance in the Leg Movements of Young Infants', *Developmental Psychology* 16 (1983), pp. 29–46.
19 An animal is decerebrate when the link between its cerebrum and the rest of its CNS has been severed.
20 P. Whelan, 'Control of Locomotion in the Decerebrate Cat', *Progress in Neurobiology* 49:5 (1996), pp. 481–515.
21 B. Hochner, 'How Nervous Systems Evolve in Relation to their Embodiment: What We Can Learn from Octopuses and Other Molluscs', *Brain, Behavior and Evolution* 82 (2013), pp. 19–30.
22 L. Zullo, G. Sumbre, C. Agnisola, T. Flash, and B. Hochner, 'Nonsomatotipic Organization of the Higher Motor Centers in Octopus', *Current Biology* 19 (2009), pp. 1632–1636.
23 G. Levy, T. Flash, and B. Hochner. 'Arm Coordination in Octopus Crawling Involves Unique Motor Control Strategies', *Current Biology* 25:9 (2015), pp. 1195–1200.
24 A.J. Ijspeert, 'A Connectionist Central Pattern Generator for the Aquatic and Terrestrial Gaits of a Simulated Salamander', *Biological Cybernetics* 84 (2000), pp. 331–348.
25 The appeal to 'animats' as examples of biological systems is contentious. I argue for the explanatory value of animat research in D. Jones, 'What Do Animat Models Model?', *The Journal of Experimental and Theoretical Artificial Intelligence* 24 (2013), pp. 475–488.
26 D. Dennett, *Consciousness Explained* (New York, NY: Back Bay Books, 1991).
27 A. Clark, 'Soft Selves and Ecological Control', in D. Spurrett, D. Ross, H. Kincaid and L. Stephens (Eds.), *Distributed Cognition and the Will* (Cambridge, MA: MIT Press, 2006), pp. 101–122, on p. 114 (original emphasis).
28 Specifically, he references the account offered in L. Wright, 'Functions', *The Philosophical Review* 82:2 (1973), pp. 139–168.
29 Burge, 'Primitive Agency', p. 255.
30 A related but distinct notion of behavior is more commonly associated with conduct: a child displays bad behavior; his arms and legs do not, and this is true even if it is the movement of his limbs that is eliciting disapproval.
31 R. Millikan, *White Queen Psychology and Other Essays for Alice* (Cambridge, MA: MIT Press, 1993), on p. 141.
32 Ibid., p. 137.
33 In M. Bickhard, 'The Dynamic Emergence of Representation', in H. Clapin, P. Staines and P. Slezak (Eds.), *Representation in Mind: New Approaches to Mental Representation* (Oxford, UK: Elsevier Inc., 2004), pp. 71–90.
34 F. Dretske, *Explaining Behavior* (Cambridge, MA: MIT Press, 1988), p. 47.
35 Millikan, *White Queen Psychology and Other Essays for Alice*, on p. 138. Millikan argues that simple computers, which can fulfill their functions without engaging the environment, do not exhibit behavior for precisely this reason.
36 Dretske, *Explaining Behavior*, p. 11.
37 Ibid., p. 26.
38 Ibid., p. 24.

4 Guideless guidance

A change of perspective

Much of our science is dedicated to understanding the relationship between parts and wholes. We frequently privilege reductive, mechanistic analyses that presume that wholes can be analyzed into their functional parts. So analyzed, we may then fully explain the behavior of wholes by appealing to intercomponential operations. This approach has been unquestionably successful, and one might attribute its success to the idea that the parts of a system are logically prior, both to the wholes they compose and to our investigation of them. From this perspective, parts are primitive components of wholes, which inherit their properties – to include their causal powers – through the operations of those parts. The task of mechanistic reduction is to explain wholes by *discovering* their parts and the relations between them.

The priority of parts is not so obvious upon reflection, however. Although it is true that wholes depend upon their parts, an object cannot be a part without being a part *of* some whole (a detached part is a former part of the whole it came from). Indeed, some parts can only be fully understood in the context of the wholes they compose. A full characterization of heart function, for example, must refer to the body of which it is a part. Moreover, it is difficult to provide criteria for an objective division of wholes into parts. How many parts does a dog have? Do we simply count the head, torso, legs, and tail? Should we divide it into subsystems (skeletomuscular, immune, reproductive, etc.)? Atoms? Quarks? Or does the dog have two parts – the front half and the back half?

We might accurately describe the dog in any of these ways, and each examination might teach us something different. Different (and overlapping) parts can be abstracted from the same whole, and equally accurate statements can be made at varying degrees of resolution. Rasmus Winther argues that the part–whole relations of a system can be studied from multiple compatible viewpoints, which he calls *partitioning frames*. Winther defines a partitioning frame as 'a set of theoretical and experimental commitments to a particular way of abstracting kinds of parts'.[1] Each frame can be distinguished by its unique norms and aims.

Winther identifies three partitioning frames in biology – mechanistic, structuralist, and historical:

1   Mechanistic explanation aims to provide an account of the whole in terms of the causal properties of basic parts. Its explananda are system behavior and development, or the mechanisms themselves.

2   Structuralist explanation intends to furnish an account of the system in terms of (i) structural parts found at many levels and (ii) the realization of formal mathematical laws in emergent processes. Its explananda are emergent system form and development.

3   The goal of historical explanation is to present a narrative, a biography, of a temporally-changing organism-part, or species qua part, by placing it in its contextual whole. Its explananda are the long-term temporal changes of organism-parts, species qua parts, and their respective wholes.[2]

Thus far I have considered explanations of primitive action offered from within the mechanistic and historical partitioning frames and found them lacking. This is not to say that the investigation thus far has been a failure – a complete study of behavior and action should include both causal accounts of the guidance mechanisms involved in behavior production and a historical account of how and why organisms came to have the mechanisms they did. But this cannot be the whole story. The problem of what makes behavior *whole-organism* must be solved within a different frame.

This is in the spirit of Winther's analysis, as he argues that distinct partitioning frames offer *complementary* accounts of the systems they are employed to study. Each partitioning frame offers a partial explanation of a system, and thus at times multiple frames must be employed to study even a single feature of a given system. On this account it is unhelpful to argue for the superiority of a reductionist or antireductionist approach: we need to understand the operations of mechanisms within a system (reductionist), but also the organizational properties of the system they compose (structuralist). The study of primitive action requires a holistic, structuralist approach in addition to the mechanistic analysis considered thus far. I will attempt to provide such an account in the chapters to come.

Chapters 2 and 3 addressed a pair of central questions for the investigation of primitive action:

1   What sort of guidance is necessary for action?
2   What does it mean for the whole organism, rather than one of its parts, to act?

Neither question was adequately answered. I now propose that they can be answered jointly by viewing agents and actions as self-organizing systems. This chapter offers a partial answer: Agents are globally organized patterns of activity. Guidance is a form of context-sensitive mutual constraint within an agential system. This is far from a complete account, but it frames the issue in a way that will allow me to advance a more detailed positive account in what follows.

Up to this point I have presumed that guidance requires a guide, much in the way that a car requires a driver. But not all orderly systems are so guided. Consider the case of a river and its riverbed. Water flowing through dirt carves out a channel, which in turn constrains the directions that the water can take. The behavior of the river is a function both of the dynamics of the water and of the structural features of the channel, and it would not be unusual to think of the channel as guiding the flow of water. But it does not guide the water through a diachronic process of monitoring and adjustment. It guides the water simply by existing as a material constraint upon the direction of its flow, which in turn changes the structure of the channel, and so on. The interdependence between the water and the channel serves as a form of continuous 'guidance through mutual constraint'. I argue in what follows that a similar (albeit more complicated) process is involved in the construction of whole-organism behavior and that the kind of interdependence characteristic of complex self-organizing systems provides the guidance necessary for agency. This form of guidance, which does not depend on reflection, deliberation, or planning, is also ideally suited for making sense of the low-level forms of guided movements that a theory of primitive agency ought to accommodate.

There are good reasons to think that human beings are complex systems. Our brains are composed of hundreds of billions of neurons, with hundreds of trillions of connections. Our immune systems are enormous networks of trillions of cells, distributed throughout the body and both moving and communicating through its blood and lymph channels. The human genome includes roughly 19,000 genes. Genetic regulatory networks arise within this system through interdependence and feedback loops whereby our genes control each other's expression. No central controller determines how these networks ought to run – their coordinated activity results from the activity of organized networks of interdependent parts. This is to say that they are *self-organizing*.

W. Ross Ashby is credited with coining the term *self-organization*. The overall state of the self-organizing system is determined by the interdependence between its parts, which Ashby calls the 'conditionality' of the system.[3] He characterizes this interdependence in terms of mutual constraint; part A constrains part B in a relation when state changes in A limit the set of possible states into which B can enter. Thus a full explanation of B's behavior must make reference to A and vice versa. In a massively interdependent complex system, a full explanation of the behavior of one part may require reference to many or even all of the other parts.

The study of self-organization has developed a great deal since Ashby's early work. Cliff Hooker lists eighteen characteristics of complex systems in his philosophical introduction to the topic.[4] Because many excellent books have already been written on complexity theory, I will not attempt a survey of the discipline in what follows. Instead I briefly summarize some of its core concepts for nonspecialists with an eye toward their application to action.

Self-organization

The deep interdependence of the self-organizing system's parts makes them behave in nonlinear fashion. The number of variables that must be accommodated in the study of even very simple self-organizing systems can make them difficult to understand without the use of advanced mathematics. Fortunately, dynamical systems theory offers resources that can aid in the visualization of complex behavior.

The behavior of dynamical systems can be represented graphically along several axes corresponding to the number of variables required to characterize the system mathematically. The possible states of the system can be represented as a region on this graph corresponding to the system's *state space*. Adding a further axis to represent time, the graph represents the system's *phase space* – the set of states into which the system may enter over time.[5] The system's actual behavior may be represented as trajectory through phase space.

Self-organizing systems are remarkable in their tendency to settle into (meta-) stable patterns of behavior. These patterns are described by the *attractors* of the system. There are many different varieties of attractor. For example, the attractor of a swinging pendulum in normal conditions is a *point attractor* – as energy dissipates from the system, it slows to a halt, and its trajectory is pulled into a single point. The pendulum can always be perturbed out of that point attractor by introducing more energy into the system (by pushing it), but its trajectory always returns to that point as its energy dissipates through friction. Other attractors are cyclic, with the system repeatedly revisiting the same states (e.g., a frictionless pendulum or a walking person). *Strange, complex, or chaotic attractors* are characterized by nonrepeating trajectories with significant fluctuation but which nonetheless stay within certain boundaries in phase space.

The phase space of a self-organizing system can thus be characterized as a probabilistic landscape of 'competing' attractors.[6] The behavior of a system is described by a trajectory through that landscape. In a completely attractor-free phase space, all regions of phase space are equally accessible to the system. Attractors describe constraints on the system's behavior – tendencies to revisit certain regions of phase space more often than others. Attractor strength is characterized by the depth of its *basin of attraction*. A system that manifests a stable oscillatory pattern, for example, may be perturbed out of that oscillation if the perturbation relocates the state of the system outside the relevant basin of attraction. Alternatively, if the system remains within the basin, it will tend to settle back into its original pattern of activity. Some basins of attraction will be larger than others – the 'deeper' the basin, the more likely the system is to be pulled into it and the more energy will be required to leave it.

For example, homeostatic processes can be characterized in terms of a 'fitness space', such as that involving temperature and heart rate in humans. Certain variations along both dimensions are permissible without disruption of the pattern; climbing a flight of stairs elevates the climber's heart rate, but when given time to rest, it returns to a baseline level. Of course, stronger fluctuations may perturb the system out of this pattern: increasing a person's heart rate beyond a certain threshold will move the system out of its viable range, at which time it will converge upon a single point attractor (corresponding to the death of the system).[7]

The nonlinearity of complex systems entails that small local changes in such systems can bring about global change in their attractor landscapes. Such changes, called *bifurcations*, can be represented as changes in the attractor landscape of a dynamical system. Rayleigh–Bénard convection offers a common example of bifurcation. Heating the bottom of a pan of cooking oil causes a temperature difference between the top and bottom of the oil layer. Initially the heat only agitates the oil's constituent particles. But once the temperature gradient (described by the *control parameter* of this system) passes a certain threshold, rolling waves form in the oil. As the value of the control parameter increases, another bifurcation occurs, and Bénard cells – hexagonal cells of rolling oil – form. The amplitude of the waves is described by an *order parameter* in the language of dynamic systems.

As the attractor landscape of a dynamical system shifts over time, its stability may vary substantially from one moment to the next. This allows very minor perturbations to drive the system into highly robust (i.e., perturbation resistant) patterns of activity. For example, when oil is heated its convection rolls may oscillate clockwise or counterclockwise. This is because as the oil is heated, system instability reaches a critical threshold and a *pitchfork bifurcation* forms (imagine a steep hill bordered by deep valleys, with the state of the system balanced precariously atop the hill). At that moment, clockwise and counterclockwise rolls of the oil are equiprobable. The matter of which way the fluid circulates is not determined by any outside organizer, but rather is settled by random perturbations to the fluid, pushing the system into one of two cyclic attractors. The basins of attraction for these attractors are extremely deep, however, so although even minor perturbations can initiate oscillation in either direction, reversing the oscillation requires a significant energy investment.[8]

It is obvious that not all self-organizing systems are agents and not all self-organized movement is action. But the previous discussion does provide us with a framework to move forward. More specifically, treating the agent as a self-organizing system allows us to understand both how the organism as a whole could produce its action and how it could be said to guide its parts in an unproblematic fashion. The human body as a whole is a massively complex system – J.A. Kelso estimates that our coordinated behavior arises from the integrated efforts of '10^2 joints, 10^3 muscles, 10^3 cell types, and 10^{14} neurons and neural connections'.[9] This integration includes not only the individual behaviors of each part, but also the incidental physical forces they exert upon one another – quickly turn your head and notice the compensatory adjustments that must be made throughout your body! Moreover, each pattern of activity is composed of innumerable subpatterns. Walter Freeman describes each pattern of cortical activity as 'a dissipative structure emergent from a microscopic fluctuation'.[10] Movements may also be viewed as self-organizing systems – not chains of details, but 'morphologically coherent and holistic forms'.[11]

It is important to note that 'pure' self-organization – in the sense of organization taking place in the absence of *any* outside influence – is impossible. The second law of thermodynamics states that the universe tends toward entropy. In light of this, it may seem remarkable that patterns of physical-state change could self-organize at all, to say nothing of the gradual increase in system complexity over evolutionary time. Given the universe's natural tendency toward disorder, why should we find increasingly ordered systems?

Appearances notwithstanding, self-organization is consistent with the second law. This is because a system must drain order from the surrounding universe to self-organize, and its persistence depends upon its continually exchanging energy or mass with its environment in such a way as to remain thermodynamically far from equilibrium, or *open*.[12] Systems that participate in this process of energy acquisition and discharge are called *dissipative* systems.[13] Deprive such a system of energy for long enough and, like the pendulum, its dynamics eventually collapse to a fixed point (this, I tell my children, is why we feed the dog).

Living systems are paradigmatic cases of open systems, but coordinated movement itself is also an open system – walking exhibits a stable cyclic pattern, but only for as long as the walker devotes the appropriate amount of energy to its sustainment. Although open systems manifest stable, ordered patterns of activity, they are never fully stable. They are better viewed as what Alicia Juarrero calls 'structures of process',[14] the metastable behavior of which fluctuates around a fixed set of values. The vital point of this section is that both agents and coordinated movements can be viewed as structures of process in this sense. The energetic requirements for their sustainment will feature centrally in the positive account to come.

Constraint as a form of guidance

The picture presented thus far depicts the agent as a dissipative, far-from-equilibrium, nonlinear, massively interdependent, complex system. This combination of interdependence and nonlinearity allows state changes to one part of a system to effect global change within that system (the nature and extent of its effect being conditional upon the states of the other parts of the system). In this section I discuss ways in which the forms of interdependence found in complex systems may produce the sort of guidance needed for primitive action.

Entrainment is a form of dynamic coupling. The seventeenth-century mathematician Christian Huygens first recorded the phenomenon, noting that two of his wall-mounted pendulum clocks would fall into antisynchrony over time. This phenomenon, which Huygens referred to as 'odd sympathy', can be explained by appealing to feedback between the two pendulums in the form of energy transfer through the wall. The feedback from each pendulum adjusts the behavior of the other until the coupled system reaches a stable global state.[15]

Similar examples have involved increasingly large numbers of randomly set metronomes placed on a platform designed to allow vibrations from each metronome to either dampen or amplify the movements of the others. In such cases the group ultimately converges upon a single rhythm. This kind of coupling, called *entrainment* or *frequency locking*, is common in nature. James Glazier and Albert Libchaber list a number of examples:

> In mechanics, the damped driven pendulum . . . in hydrodynamics, the vortices behind an obstacle in a wind tunnel or an airplane wing . . ., the dripping of a faucet, the convective rolls in a pan of heated water, and the oscillations of acoustically driven helium . . .; in chemistry, the Belousov-Zhabotinsky reaction, the Chlorite-Thiosulphate reaction and many others . . .; in solid

state physics, charge density waves in niobium selenide . . ., and other com-
pounds . . ., the conductivity of barium sodium niobate . . ., oscillations in
Josephson junctions . . ., and in germanium . . ., in biology, cardiac cells . . .,
the brain . . ., the slime mold dyctiostelium discoideum . . ., menstrual cycles
in human females, and elsewhere. . . . This list could probably be extended
almost indefinitely.[16]

The fact that this phenomenon manifests in systems that have virtually nothing
else in common is particularly interesting. It is a feature of certain relational organizations rather than any particular sort of matter. Feedback between the elements
of each system produces changes in each element without the presence of a direct-
ing influence. The result is order at a global level. This serves as the most basic
form of coordinator-free coordination.

Entrainment is a form of constraint. Juarrero distinguishes two forms of constraint: context sensitive and context free. Context-free constraint affects *prior*
probabilities within a system: for example, a weighted die might roll a six more
often than a fair die.[17] Context-free constraints may push systems out of equiprob-
ability but they do not generate complexity – to use Juarrero's example, one cannot build a sand castle simply by piling more sand onto a heap. If constraints only
limited the permutations that a system could undergo, they would not be able to
explain the complexity of self-organizing systems.

Juarrero argues that complexity arises from the imposition of context-sensitive
constraints on systems. Unlike context-free constraints, context-sensitive constraints
affect the *conditional* probabilities within a system. They obtain when previously
independent parts become interdependent, forming a system that in turn functions
as the context within which the behaviors of those parts are correlated.

Consider the laser: Prior to excitation its constituent atoms exhibit a super-
position of modes of activity, emitting individual light waves of random phase.
Increasing the amount of energy in the system strengthens these competing patterns, and those with the highest rates of increase come to dominate the weaker
patterns. Upon passing a certain energetic threshold, one pattern of activity – 'a
practically infinitely long sinusoidal wave' – gradually dominates all others,
thereby restricting the freedom of the local light particles. Those particles regain
their freedom of movement only when a sufficient amount of energy dissipates
from the system, collapsing the wave. Hermann Haken described this process
both as 'cooperative' and, more ominously, as *enslavement*.[18]

Haken's description of the laser's self-organization as 'cooperative' is sugges-
tive. Much as members of a community might agree to give up certain personal
freedoms to thrive as a collective, the complex system acquires new capacities
through the mutual restriction of its parts. If some cosmic accident caused my
constituent atoms to separate into a free-floating mist, each atom would enjoy
considerably more freedom than it had before the accident. But the likelihood that
the collective would be able to type this sentence, or do anything else of interest,
is incredibly low. It is only when they are bound into a complex, organized whole
that the whole is able to type. In this sense context-sensitive constraint is enabling.

Guided movement can then be understood as a case of a system allocating energy to and imposing constraints on its constituent parts to generate the right patterns of self-organized activity at the right times – fueling structures of process.

Self-organized behavior

The Russian physiologist Nicolai Bernstein viewed the cognitive task of behavior production as one of reducing redundancy in a system with an enormous number of degrees of freedom. The embodied agent solves this problem by grouping its elements into unified wholes, that is, reducing vast multivariable systems to more controllable ones governed by a few order parameters. Kelso notes that this process is analogous to *i*-tuple formation in language: just as interdependence in letter frequency leads to increasingly large clusters of letters with increased expressive power, interdependent neural impulses and physical movements produce clusters of movement called *synergies*:

> [S]ynergies constituted a dictionary of movements in which the efforts of the muscles were the letters of the language, and synergies combined these letters into words, the number of which was much less than the number of combinations of letters . . . the language of synergies was not just the external language of movements but also the internal language of the nervous system during the control of movement.[19]

Synergetics treats movement as a process of constructing a coherent functional form from multiple parts. This process is partly determined by body morphology (the wrist and elbow joints always will work as a unit – a context-free constraint imposed by the radius and ulna linking them), but far more interesting behavioral units can be constructed. As noted earlier, synergies are open structures of process– they require energy both for their construction and for their sustainment over time.

The same movement may be assembled in quite different ways. Thelen and Smith discuss the case of motor control in two very different infants. The first, Gabriel, was extremely active and prone to high-velocity arm flapping movements. Thelen and Smith note that Gabriel's flapping had the same dynamical characteristics as those of a spring or pendulum; his arm velocity varied smoothly with lateral displacement, indicating a cyclic attractor. Gabriel's reaching required 'massive co-contraction of both shoulder and upper arm movements', which effectively stiffened the arm. This damped the flapping movement and channeled it toward a single target location. Or, as Thelen and Smith put it, he converted 'a cycle attractor pattern into a point attractor, much as a naturally damped spring reaches a point equilibrium'.[20]

The second child, Hannah, was quiet and less active overall. Hanna's movements were slower and more deliberate; she stiffened her arm enough to lift it from her lap but no more than that, and the bulk of her force was devoted to overcoming gravity rather than her own activity.[21] These cases show that a single movement can be assembled from very different components. But the reaching movements shared an important similarity: they both possessed the ability to channel their internal dynamics into successful movement. Both children solved the coordination problem of reaching in ways distinct from those employed by mature adults. This suggests that development may be viewed as a process whereby old solutions to coordination problems are destabilized and dissolved and new solutions emerge.

Thelen and Smith's analysis of child reaching is but one of many dynamical analyses of coordinated behavior, and such analyses frequently identify patterns common to various complex systems. J.A. Kelso's study of interlimb coordination is one famous example.[22] Kelso's subjects wagged their index fingers rhythmically back and forth in two stable patterns: in phase (simultaneously activating homologous muscle groups) and antiphase (alternately activating those groups). They were then prompted to slowly increase their rate of movement. Subjects who started the task with antiphase movement would spontaneously switch to in-phase movement. Subjects who started in-phase would remain there. Thus it seems that the coordinated system has two stable states up to a certain threshold, at which time the number of possible stable states collapses to one – a bifurcation. Equally curious is the fact that subjects who were subsequently asked to slow their rate of movement took slightly longer to return to antiphase movement. This phenomenon, known as hysteresis, is characteristic of nonlinear complex systems.

The model of this coordinated behavior, known as the Haken-Kelso-Bunz (HKB) model, governs other activities, to include finger wagging in split-brain patients[23] and synchronized leg swinging in groups.[24] As in other complex systems, feedback between elements of the system is necessary: The swinging was not coordinated when the subjects were not looking at each other, but the introduction of visual feedback caused the coupled system to manifest the same pattern of activity described by HKB. The model appears to be more than a description of a particular mechanism, but rather a general principle of pattern formation. This provides further support for the idea that principles of self-organization found in even simple autocatalytic systems may also be applicable to the behavior of whole-organism systems.

This offers a partial alternative solution to the problems of coordination and guidance discussed thus far. Coordination *is* necessary for action, but it is not imposed by a coordinating mechanism. It arises from the sort of entrainment, slaving, and synergy-building characteristic of self-organizing systems. The agent is a self-organizing system comprising various other self-organizing systems, to include not only its parts but also its movements. The agent cannot be separated from its movement as its guide or author any more than the laser can be separated from the sinusoidal organization of its parts. Rather, the agent supervenes on the networks of self-organizing processes that constitute its movements. Thus Wittgenstein's question – what is left over when we subtract my arm's going up from my raising my arm? – is a trick question. The subtraction is not so easily performed.

This way of viewing the relationship between agent and action also allows us to solve the nanodeviance problem raised in Chapter 2. Recall that the problem was that some disqualifying source of causal deviance could always be inserted into a diachronic, efficient-causal relation between the guiding system and the behavior it was intended to guide. My guidance system might prescribe that I raise my arm, but the message will only get to the relevant subsystems if the hijacking agency allows it. Thus the action is no longer my own. So, the prescriptions of a guidance system are insufficient for action.

Things are quite different if the guidance relation is understood as a form of contextual constraint among interdependent parts. Here the behavior will count as action if it is a synergy assembled from the complex of interdependent parts of that constitute the agent and is subordinated to agent-level patterns of activity. To determine whether this criterion is met, we check the dependence relationships that obtain among the parts. We see in the case of nanodeviance that the dependence relationships are *not* confined to the constituent parts of the agent – the chain of control extends outside the agent to include the operations of the neuroscientist. The resulting behavior is properly attributed to the agent–nanocontroller–neuroscientist system, not the agent, and thus fails as a case of (the agent's) action.

Contrast this with the case of an implanted chip that relays incoming information *only* to the intended receiving station. Here the dependence relations are kept 'in-network', so to speak, and the agent will generate the relevant patterns of activity through the mutual constraint of its parts. The agent is no mere internal controller relaying commands to its body, but is rather a process that supervenes on its own behavior, much as the sinusoidal laser wave supervenes on the behaviors of its enslaved light particles. The *synchronic* form of coordination manifested by self-organizing systems is the missing ingredient for action sought in Chapter 2.

Here new and interesting problems arise, however. I have suggested that agents are self-organizing systems. But agents are also composed of self-organizing systems. Their movements are self-organized synergies. Moreover, agents are embedded in dynamical environments. Looking at this world of patterns within patterns, it is far from obvious when a given process is subordinated to an *agential* system rather than some other complex system. Self-organization, although necessary for the kind of guidance characteristic of agency, is insufficient – we still need ways to distinguish genuine cognitive agents from other self-organizing systems. This is to say that we still have no principled way of determining when the agent as a whole, rather than one of its parts, is producing an action. Juarrero suggests that agent-location amounts to identifying the proper collective variables in a complex system:

You and I (as well as tornadoes and slime molds) are . . . local eddies of order. Because these processes are local eddies of order in a sea of much higher dimensionality and potential, fewer dimensions are required to describe them; self-organized processes are therefore describable as collective variables: you and I. Folk psychological terms are just that: descriptions of the collective variable level.[25]

The problem is that it is far from obvious what should guide the process of order parameter selection. Indeed, from the standpoint of the complexity theorist, the matter of what systems count as agents and which do not may not be settled objectively – in a world of flux, boundaries may be drawn as the observer sees fit. But this threatens the possibility of an objective notion of agency, as well as the possibility of an objective distinction between agent and environment. I elaborate the challenge in what follows.

Mild pattern realism and fuzzy boundaries

Self-organizing systems are patterns of state change. In adopting a systems perspective on action, we commit ourselves to taking those patterns seriously (rather than as mere epiphenomena of their parts), but in so doing we risk losing a characterization of agency as observer independent. Daniel Dennett suggests that patterns are, by definition, 'candidates for pattern recognition', where this is supposed to imply a 'loose but unbreakable link to observers or perspectives'.[26] This is not to deny the reality of patterns, but only to suggest that there is no observer-independent fact of the matter as to which patterns are the *real ones*. The question of whether some pattern is real, for Dennett, is settled by whether '*there is* a description of the data that is more efficient than the bit map, whether or not anyone can concoct it'.[27] Multiple – and even conflicting – patterns can thus be discerned from nearly any data set, and all are equally real by Dennett's criteria. The fact that we frequently see the same patterns in interpreting our environments, Dennett suggests, is due to what Sellars called our shared 'manifest image' of the world, which Dennett takes to result from a combination of pragmatic and evolutionary considerations. The fact that humans tend to see the same pattern in a particular case demonstrates very little about the observed data themselves.[28]

If Dennett is correct, there might be no objective fact about whether some movement counts as an action or whether some system counts as an agent, in which case the subjectivist specter looms yet again. Indeed, from this perspective it seems particularly dangerous – the most full-blooded intentional action might merely be intentional 'for us', as one can imagine the case of an alien with radically different ways of engaging the world examining a human being at multiple levels and finding nothing but disorganized noise. If Dennett is right, the alien's perspective is no less correct than our own. We happen to be wired to view one another as agents, but there is no further fact of the matter.

Similarly, Ashby notes that the kind of organization seen in a system is largely observer dependent, but also suggests that for this reason the boundaries of a given organization may be applied arbitrarily, sometimes generating counterintuitive results, such that 'any dynamic system can be made to display a variety of arbitrarily assigned "parts", simply by a change in the observer's viewpoint'[29] and, more radically:

Take a dynamic system whose laws are unchanging and single-valued, and whose size is so large that after it has gone to an equilibrium that involves only a small fraction of its total states, this small fraction is still large enough to allow room for a good deal of change and behavior . . . then examine the equilibrium in detail. You will find that the states or forms now in being are peculiarly able to survive against the changes induced by the laws. Split the equilibrium in two, call one part 'organism' and the other part 'environment': You will find that this 'organism' is peculiarly able to survive against the disturbances from this 'environment'. . . . thus, as I said, every isolated determinate dynamic system will develop organisms that are adapted to their environments.[30]

The latter quote demonstrates both strengths and weaknesses of Ashby's conception of self-organization. On one hand, it requires us to view systems as embedded; their stability both arises from and depends essentially upon the variables governing the state of their surroundings. Ashby's commitment to embedding can also be seen in his rejection of a sense of 'self-organization' that suggests that a system could change the laws governing its state transitions over time (laws which are encoded in the functional mapping that defines the system). Adaptive behavior is not the result of a complex system's reorganizing its structure in a vacuum, but rather by the dynamic interplay between a deterministic system and a complex environment.

Thus the appearance of being 'self-organizing' can be given only by the machine S being coupled to another machine (α) . . . Then the part S can be 'self-organizing' within the whole S + α. Only in this partial and strictly qualified sense can we understand that a system is 'self-organizing' without being self-contradictory.[31]

This supports a deeply embedded conception of systemhood. Self-organizing systems are self-assembling systems, and the behaviors of those systems can change over time in response to changes in environment. But we should not view these changes as system-driven organization changes. Rather, each system should be viewed as a pocket of equilibrium within a larger system; it exists insofar as it has formed under these conditions, and it persists in maintaining its equilibrium in those conditions. Its 'behavior' depends essentially upon the interplay between the system and its environment; the two cannot be separated without loss of understanding. This motivates (but does not entail) the thesis that there are no hard-and-fast distinctions between agents and environments – they constitute parts of a larger complex system and can be considered as distinct only for the purposes of some inquiry or other. Moreover, if intelligent behavior can only be viewed as a property of the coupled agent/environment system, then it becomes attractive to view the boundary between agent and environment as artificial.

It seems that by embracing complexity theory we find ourselves awash with patterns and no principled means of distinguishing the agential patterns from the others. The present account could, following Dennett's suggestion, proceed as a Sellarsian 'synoptic view' integrating the manifest and scientific images. But I believe a stronger position should be taken. Certain organizations of matter distinguish *themselves* as essentially agential, and they do so independently of outside observation. Agents are specialized kinds of complex systems. In the next chapter I explore the possibility of grounding agency in the operations of living, self-producing systems.

Top-down causation and emergence

When discussing self-organized systems, there is constant temptation to describe the higher-order properties of the system as something distinct from their constituent parts. Even in careful dynamical analyses one finds talk of order parameters *leading* systems through state changes or attractors *pulling* the system into different configurations. This habit is no doubt motivated by the qualitative nature of dynamical systems analyses, and it is convenient – perhaps indispensible – for describing the global patterns that form in such systems. But it is a loose way of speaking – order parameters are components of mathematical *descriptions* of patterns. Attractors describe changes in the likelihood that the global system will fall into certain patterns. They are not additional causes – what manner of force could an order parameter exert?

Many complexity theorists describe these higher-order systemic patterns as *emergent* entities that apply *top-down causes* to their parts. These terms are frequently used in the more benign epistemic sense that the nonlinear dynamics of complex systems render them impossible to understand at the level of local interactions. But some take the relationship between global pattern and local interaction to be a form of interlevel *causation*: once the whole has been produced, it exists as a new *entity* that exerts new causal *forces* upon its parts. For example, Juarrero argues that the contextual constraints that hold between parts of a self-organizing system give rise to emergent wholes, which in turn exert second-order contextual constraints on their parts.[32] The process of enslavement would seem to be one such process – once the parts of a system form the whole ('causing' it to 'emerge'), the emergent pattern of activity binds them into its continued service, at least until energy levels dip below whatever threshold was needed to sustain it.

These more metaphysically robust notions of emergence and top-down causation, if defensible, would benefit the present project. But both notions are contentious, and neither is required for a complex systems perspective or the theory of agency it informs. Indeed, there is good reason to think that at any given time the state of a complex system *can* be explained in terms of interactions – to include first-order contextual constraints – between its parts. The fact that a part contributes to the sustainment of a global pattern of activity does not establish that the global pattern is causing it to do so. The deep interdependence of these parts, however, does make reference to a whole network necessary in many cases. If each part of a system depends on the others, then to explain the behavior of one part, we must explain what all of the parts in that system are doing, and in this case talk of global patterns of activity becomes indispensable. We can also talk about the activities of parts being subordinated to the activities of wholes in the sense that once the dependence network reaches a certain size, new constraints are imposed upon local interactions. But none of this requires the addition of new entities or causal forces.

This is not relapse into antirealism, as though we arbitrarily pick epiphenomenal patterns in describing agent-guided behavior. Systems are defined by objective facts about their organizations. For example, certain self-organized systems are operationally *closed* in such a way that the dependence relations obtaining between their parts exhibit a self-contained, recursive structure – the system, understood as a process, feeds itself in ways that distinguish it from its surroundings as a whole. In so doing the system defines the environment in which it operates – the ground against which it stands as figure. Such systems also have characteristic stability requirements – the requisite energetic input for the sustainment of convection rolls is quite different from (and indeed, inconsistent with) that required for the sustainment of a goldfish.

Second, one can consistently deny metaphysically robust accounts of top-down causation and emergence *and* claim that the mechanistic perspective only offers a partial account of reality. A process-based, dynamical perspective is needed to make sense of the unfolding of systemic behavior over time. In the next chapter I will show that living organisms are essentially temporally extended systems that cannot be identified with their material configuration at any given time. They distinguish themselves as wholes by maintaining invariant organizations through processes of material change. The demand to maintain this organization constitutes the earliest form of system-level normativity. These processes require a systems perspective within which talk of *the whole organism* is necessary.

We will see that the agent is a closed, autonomous system that both contains and participates in numerous other closed, autonomous systems. As a result, the same part may participate in the activities of numerous wholes. The question of whether a given pattern of activity counts as action will be settled by determining to which system it contributes during its operation and how. The next two chapters will flesh out what distinguishes agents from other complex systems and what it is for a process to be governed by its dynamics.

Notes

1 R. Winther, 'Part-Whole Science', *Synthese* 178 (2011), pp. 397–427, on p. 401.
2 Ibid., p. 408.
3 W.R. Ashby, 'Principles of the Self-Organizing System', in H. Von Foerster and G.W. Zoph, Jr. (Eds.), *Principles of self-organization: Transactions of the University of Illinois Symposium* (London, UK: Pergamon Press, 1962), pp. 255–278, on p. 266.
4 C. Hooker, 'Introduction to the Philosophy of Complex Systems', in C. Hooker (Ed.), *Handbook of the Philosophy of Science, Volume 10: Philosophy of Complex Systems* (Waltham, MA: Elsevier B.V., 2011).
5 This distinction between state space and phase space is made by Juarrero (1999), p. 152. However, it is common for complexity theorists to use the terms interchangeably. Nothing in what follows depends upon this dispute.
6 There is a natural tendency when viewing dynamical landscapes to think of attractors as 'pulling' the trajectory of the system in various ways. For example, Juarrero

(1999) describes attractors as 'representations of natural precursors of final cause', noting that 'etymologically, even the word "attractor"' suggests a pull' (p. 152). Juarrero argues at length for the re-establishment of Aristotelian final causes in modern science, and in light of her discussion, it seems plausible to think of final causation in terms of a system's tendency to visit certain regions of phase space. But no such causation is *entailed* by the dynamicist approach – a more conservative probabilistic description of the attractor landscape is sufficient for understanding the concepts as they are used here.

7 E. Thelen, 'Self-organization in Developmental Processes: Can Systems Approaches Work?', *Systems in Development: The Minnesota Symposia in Child Psychology* 22 (1989), pp. 77–117.

8 It is notable that thinking of agency in these terms highlights the absurdity of what have come to be known as Buridan cases – hypothetical scenarios where an agent is placed between two equally desirable objects and, being unable to choose either over the other, chooses neither. The twelfth-century Islamic philosopher al-Ghazali argued that only free will could settle the matter (Kane 2005, p. 37). Buridan (1340) argued that no rational choice could be made, and Spinoza suggested in his Ethics that anyone who found themselves in such a situation would behave irrationally. Reflection on the phenomenon of symmetry breaking deflates the Buridan case in a different way: the case amounts to a phase portrait of a system balanced precariously between two basins of attraction, whereupon a small nudge in either direction – perhaps a shift in balance or an attention-grabbing change in the environment – would cause the system to collapse into a stable pattern of activity. Because such fluctuations are common in any real-world system, we can see that Buridan cases can be set aside without having to appeal to notions of free will or rationality.

9 J.A. Kelso, *Dynamic Patterns: The Self-Organization of Brain and Behaviour* (Cambridge, MA: MIT Press, 1995), p. 37.

10 W.J. Freeman, *Societies of Brains: A Study in the Neuroscience of Love and Hate* (Hillsdale, NJ: Erlbaum, 1995), p. 51.

11 E. Thelen and L.B. Smith, *A Dynamic Systems Approach to the Development of Cognition and Action* (Cambridge, MA: MIT Press, 1994), p. 76.

12 I. Prigogine and I. Stengers, *Order Out of Chaos* (New York, NY: Bantam).

13 Kelso (1995) provides a slightly different characterization of dissipative systems, defining them as systems for which energy 'doesn't diffuse uniformly but is concentrated into structural flows that transport the heat (dissipate it) more efficiently' (p. 16). I choose Prigogine and Stengers' notion for its simplicity and because the features Kelso is highlighting will be mentioned in the upcoming discussion of nonlinearity and bifurcation.

14 A. Juarrero, *Dynamics in Action: Intentional Behavior as a Complex System* (Cambridge, MA: MIT Press, 1999), p. 124.

15 E. Klarreich, 'Huygen's Clocks Revisited', *American Scientist* 90:4 (July–August 2002), at www.americanscientist.org/issues/pub/huygenss-clocks-revisited [accessed 15 March 2014].

16 J. Glazier and A. Libchaber, 'Quasi-periodicity and Dynamical Systems: An Experimentalist's View', *IEEE Transactions on Circuits and Systems* 35:7 (1998), on p. 790.

17 A. Juarrero, *Dynamics in Action: Intentional Behavior as a Complex System* (Cambridge, MA: MIT Press, 1999), pp. 134–136.

18 H. Haken, 'Synergetics: An Approach to Self-Organization', in E. Yates (Ed.), *Self-Organizing Systems: The Emergence of Order* (New York: Plenum Press, 1987), pp. 417–437, on p. 420.

19 Ibid., pp. 38–39.

20 Thelen and Smith, *A Dynamic Systems Approach to the Development of Cognition and Action*, p. 259.

21 Ibid., p. 260.

22 J.A. Kelso, 'On the Oscillatory Basis of Movement', *Bulletin of the Psychonomic Society* 18 (1981), p .63.
23 B. Tuller and J.A.S. Kelso, 'Environmentally-specified Patterns of Movement Coordination in Normal and Split-brain Patients, *Experimental Brain Research* 75, pp. 306–316.
24 R.C. Schmidt, C. Carello, and M.T. Turvey, 'Phase Transitions and Critical Fluctuations in the Visual Coordination of Rhythmic Movement between People', *Journal of Experimental Psychology: Human Perception and Performance* 16 (1990), pp. 227–247.
25 A. Juarrero, *Dynamics in Action: Intentional Behavior as a Complex System* (Cambridge, MA: MIT Press, 1999), p. 145.
26 D. Dennett, 'Real Patterns', *The Journal of Philosophy* 99:1 (1991), pp. 27–51, on p. 32.
27 Ibid., p. 34. Original emphasis.
28 Dennett notes that whereas insect-eating birds tend to pinpoint individual ants when feeding, anteaters tend to average out ant locations and aim for the center of an infested region. As Dennett put it, 'one might say that, while the bird's manifest image quantifies over insects, 'ant' is a mass term for anteaters' (Ibid, p. 36 note 15).
29 W.R. Ashby, 'Principles of the Self-Organizing System', pp. 255–278, on p. 260.
30 Ibid., p. 272.
31 Ibid., p. 269.
32 Juarrero, *Dynamics in Action: Intentional Behavior as a Complex System*, p. 145.

5 From eddies of order to wellsprings of value

In search of the agent

The shift to a complex systems approach to agency presents new challenges. Agents are complex systems, but they are also composed of other self-organizing systems. Their movements are self-organized, morphologically coherent patterns of state change. They operate as parts of larger complex systems (niches, eco-systems). Thus it will be necessary to offer a principled means of distinguishing agents from both the systems they comprise and the systems they compose. More broadly, self-organization might be the right form of guidance, but it remains to be determined when it is the agent that is doing the guiding. Call this the problem of self-determination.

A related problem concerns the normative, purposeful character of action. If the task of agent identification amounts to merely picking patterns out of flux, then it seems to be up to the observers of a system to determine whether a given pattern is operating *in order to* do something. Observers would assign the teleology of active behavior; purpose would not be intrinsic to the observed systems. But this denies the distinct phenomenology of agency discussed in Chapter 1 – my attention is immediately drawn to a failed action because it has failed, not because I decide upon consideration to interpret it as a failure, and certainly not because an outside observer has so decided it. Call this the problem of value.

Hans Jonas recognizes the problem of value in his criticism of the cybernetic movement of the mid-twentieth century. Jonas compares two cybernetic systems: a human and a machine, each of which has failed a task. Upon failing,

> [t]he patient finds his inability to perform distressing. But the machine, for all we know, may just as well be said instead of being distressed, to abandon itself with relish to its wild oscillations, and instead of suffering the frustration of failure, to enjoy the unchecked fulfillment of its impulses. 'Just as well' amounts of course to 'neither'. . . In the case of the machine 'missing the goal' means, of course, missing *our* goal, the goal for which it has been designed.[1]

Jonas' point is that the artifact lacks an essentially normative perspective on its behavior and that this point of view may be a necessary component for genuine agency of any sort. The machine's purposes are those of its designer. The human's purposes are self-generated. Put slightly differently, true agents exhibit original, rather than derived, purpose. I will show in this chapter that the solutions to the problems of self-determination and value share a common origin. The foundations of agency are located in the organizationally closed, autonomous processes of self-production that constitute living systems.

Metabolic freedom

Jonas offers a distinctive account of the place of life in the universe, identifying the living organism as an autonomous locus of value. This value stems from the 'needful freedom' of the organism, which is engendered by the very metabolic processes that differentiate it from its environment. In this section I provide an account of how complex systems can objectively distinguish *themselves* from their environments.

Jonas characterizes the historical development of organismic life as a series of increasingly complex systems that have consequently enjoyed progressively more freedom of action from their environment. He defines the concept of freedom as a uniquely organic 'manner of executing existence':

> [I]t is in the dark stirrings of primeval organic substance that a principle of freedom shines forth for the first time within the vast necessity of the physical universe – a principle foreign to suns, planets and atoms.[2]

Jonas's unusual application of the term 'freedom' to the lowest levels of organic life is due to his belief that free action as typically conceived has its roots in these basic metabolic processes, which are distinct from those governing other kinds of self-organizing systems (such as solar systems and atoms). His claim that living systems *execute* their existence rather than simply existing suggests that life is an active process, that is, a process that requires physical work to sustain itself. The first form of freedom arises when 'identity is wrested in a supreme, protracted effort' from the dead matter of the universe. This 'break-through of being' can be viewed as a description of the self-organizing system's apparent defiance of the second law discussed in the previous chapter – the free system first and foremost 'oppos[es] in its internal autonomy the entropy rule of general causality'.[3] The life of an organism is characterized by a continuous pressure to define and preserve its form against an increasingly entropic, lifeless background. This demand is not externally assigned in the way that the function of an artifact is assigned by its designer. It is a condition on the very existence of the living system.

Jonas argues that living organisms are unique systems in that they are essentially concerned with their own production. Through the process of self-production the organism differentiates itself as an objective individual, or unity:

> In living things, nature springs an ontological surprise . . . an entirely new possibility of being: *systems of matter that are unities of a manifold*, not in

virtue of a synthesizing perception whose object they happen to be, nor by the mere concurrence of the forces that bind their parts together, but in virtue of themselves, for the sake of themselves, and continually sustained by themselves . . . *This active self-integration of life alone gives substance to the term 'individual': it alone yields the ontological concept of an individual as against a merely phenomenological one.*[4]

Jonas believes that the unity of a living system lies not in its being a candidate for pattern recognition, but rather in its own organization, the 'active self-integration' of which involves not only self-organization and self-sustainment, but also purpose (their integration is carried out 'for the sake of themselves'). The uniqueness of the living system arises from the role that metabolic processes of self-generation play in its organization.

In Chapter 4 I noted that living systems, like all thermodynamically open, self-organizing systems require energetic input to maintain their characteristic dynamics. But Jonas cautions us against the view that the significance of metabolism lies solely in its ability to add energy to a complex system. A car engine requires fuel to operate, but its parts do not depend on fuel for their existence, nor does the engine dissolve if it runs out of fuel. Moreover, the function of the engine is to propel an object (or, tellingly, to fulfill whatever function its designer has assigned to it). It does not function to replace, rebuild, or maintain its own parts.[5] By contrast, the totality of the living system participates in the process of self-production: metabolism is essential to the persistence of not only the organism, but each part of the organism as well, down to the level of the living cell.[6]

The centrality of metabolic self-production to the living system distinguishes organismic life from not only machines, but also many naturally occurring self-organizing systems. Crystals, for example, grow by a self-amplifying process of nucleation whereby attractive forces between solute molecules bind them together into a proto-crystal, or nucleation site. The site exerts increasing pull over surrounding molecules, particularly at 'rough' edges of the crystal. Over time solute molecules fill the gaps in the rough edges, smoothing them out and slowing the rate of aggregation. The final state of the crystal is a stable lattice characterized by a three-dimensional crystallographic 'space group'.[7] It exists as a static entity – it self-organizes but does not self-produce.

Note that although crystals do self-organize, their organizations can be distinguished from the processes by which they organize. For example, suppose that a cosmic hiccup led to the instantaneous generation of a crystal lattice, which existed for a single moment before its equally abrupt destruction. However improbable, this scenario is coherent because the lattice can be fully realized at a single instant. Actual crystals have distinct temporal parts but do not need them as a condition of their existence. Moreover, nothing about the essence of a crystal requires that it grow or decay to remain as such. A crystal lattice that stops growing remains a crystal. Persistence of material suffices for persistence of crystalline identity.

The same is not true of living systems, which cannot be identified with *any* particular configuration of matter. The persistence of the living system as a unified whole depends on its taking part in continuous processes of catabolism (breaking down molecules to obtain energy) and anabolism (synthesizing nutrients).[8] Thus living systems are by definition temporally extended processes of change. If any two adjoining time slices of an organism are qualitatively identical, the system they compose is no longer living, but either dead or dormant. The predicate 'is alive' picks out a very different sort of property than predicates like 'is heavy', 'is blue', or 'is spherical'. The essential properties of a heavy blue sphere are fully realized at a single moment in time. The essential properties of a living organism, by contrast, are temporally extended: the complete physical state of the living organism at a single moment in time is but a temporal part of the organism, not the organism itself.

For this reason the mere persistence of material is insufficient for persistence of identity in living systems. Nor can the organism be identified with the sum of the material configurations that it realizes over the course of its life without indexing each configuration to a particular time in that organism's history. We can see this by imagining the life of the organism collapsed along the temporal dimension – imagine a series of static structures, each a physical configuration that an organism would have instantiated at each time step of its life, laid end to end. The result would not be a living organism but a collection of dead matter.

To say that the living organism cannot be identified with its constituent material at a given time is not to espouse dualism or vitalism, but rather a commitment to the ontological reality of structures of process. Living systems can be characterized in terms of *organizations* that are held invariant throughout the process of material change. Unlike the structure of a particular system, which consists of the sum of relations holding between its actual parts, the organization of a system consists of the set of relations that designate it as a member of a class of entities. For example, the organization of a crystal is the set of relations between parts that define them as crystalline. It is also what defines different kinds of crystals as members of the same class of objects. The discussion thus far indicates that the organization of the living involves self-production, but this claim must be elaborated and supplemented with additional properties to see how the organization of a living system could guarantee the sort of freedom that Jonas ascribes to it.

Operational and organizational closure

The first essential element of organismic organization – operational closure – is shared by many self-organizing systems. A system is operationally closed when:

> For any given process P that forms part of the system (1) we can find among its enabling conditions other processes that make up the system and (2) we can find other processes in the system that depend on P.[9]

This definition is similar to the mathematical sense of operational closure, where operations performed on members of a set yield members of that set. Crucially, to say that a system is operationally closed is *not* to say that it is closed off from interactions with its environment, but rather to say that its operations are circular and recursive – all of the operations within the system have an effect within that system, and the causes of each component's operations can be traced to other components in the system. But operationally closed systems can (and in many cases, *must*) exchange matter and energy with their environments. We might say that such systems are operationally closed but structurally open.

'Operational closure' is sometimes used interchangeably with 'organizational closure',[10] but several authors have usefully distinguished the two processes.[11] I use the term 'operational closure' to refer to the recurrent activity of parts in a closed network of interdependent processes. Operationally closed systems include living systems as well as perceptuomotor feedback loops, Bénard cells, bird flocks, and phase-locked metronomes. The characteristic circularity of an organizationally closed system, by contrast, involves those relations involved in the self-production of a system. Evan Thompson describes this process as 'dynamic co-emergence': a process through which the parts of a system coordinate to produce a unified system which in turn functions to produce its parts and sustain their operations.[12]

Operational closure is necessary, but not sufficient, for organizational closure. Perceptuomotor loops, for example, are operationally closed – successful movement requires constant feedback loops between perceptual and motor systems over the course of its construction – but not organizationally closed. Indeed, the success of a reaching pattern depends on the eventual completion of its constituent pattern of activity, not on its continued persistence or renewal. By contrast, the living cell is essentially 'a factory that makes itself from within'.[13]

Autocatalysis is one of the most basic forms of organizational closure. The products of autocatalytic chemical reactions serve as inputs to those very reactions. Once the causal loop closes, a process 'becomes a focus of influence, a self-organized eddy that draws matter and energy into itself' to maintain its organization over time.[14] The system's circular organization distinguishes it as an operationally self-contained whole against the background of its environment. This invariance of form over time can be seen as the first kind of system-determined individuality.

Autocatalytic systems are not agents. Although the autocatalytic reaction is an objective 'eddy of order', its behavior is not sufficiently independent from its surroundings to qualify it as genuinely agential. But the general notion of organizational closure can be elaborated. Here the concepts of autonomy and autopoiesis can bring us a step closer to agency.

Autonomy and autopoiesis

The concepts of closure and autonomy are not always clearly distinguished, but autonomous systems are typically specialized cases of operationally closed systems that are capable of regulating their interactions with the external environment. This is often described as a form of self-governance. More specific characterizations vary. Autonomous robots, for example, are simply capable of performing tasks in the absence of direction. Although this form of autonomy does require operational closure, most autonomous robots are not concerned with generating their own components. Other definitions of autonomy embrace organizational closure. For example, Kepa Ruiz-Mirazo and Alvaro Moreno define 'basic autonomy' as

> the capacity of a system to manage the flow of matter and energy through it so that it can, at the same time, regulate, modify, and control: (i) internal self-constructive processes and (ii) processes of exchange with the environment.[15]

Cliff Hooker defines it as

> the internally organized capacity to acquire ordered free energy from the environment and direct it to replenish dissipated cellular structures, repair or avoid damage, and to actively regulate the directing organization so as to sustain the very processes that accomplish these tasks.[16]

Both definitions presuppose that the autonomous system is a far-from-equilibrium, organizationally closed, dissipative, self-producing system. Both also employ normative language: the autonomous system collects and manages the flow of matter and energy *in order* to maintain its organization and successfully cope with its world. This abstract characterization can be made more concrete by examining a case of biological autonomy – the autopoietic system.

The Chilean biologists Humberto Maturana and Franscisco Varela define the organization of the living system in terms of an autopoietic machine:

> An autopoietic machine is a machine organized (defined as a unity) as a network of processes of production (transformation and destruction) of components that produces the components which: (i) through their interactions and transformations continuously regenerate and realize the network of processes (relations) that produced them; and (ii) constitute it (the machine) as a concrete unity in the space in which they (the components) exist by specifying the topological domain of its realization as such a network.[17]

This circular process amounts to the self-production of the system. Autopoietic systems can be distinguished from allopoietic systems, which do not produce their own components, and heteropoietic systems, which are the product of human artifice.[18] The autopoietic model distinguishes the uniqueness of living systems in the physical world without vitalist appeals to unique substances.

Varela later emphasized the role of a spatial boundary in a simplified set of criteria for autopoiesis: the autopoietic system has a semipermeable boundary separating its internal processes from its environment, that boundary must be produced from within the system, and it must encompass reactions that serve to regenerate components of that system.[19] The boundary condition distinguishes the autopoietic system from other organizationally closed systems. In the cell, this boundary takes the form of the cell membrane. The boundary is both the product of and a necessary condition for the cell's metabolism: having been formed by metabolic processes, it admits only materials from its environment necessary for its persistence (when all goes well). The boundary both limits and takes part in the internal dynamics of the system, whereas the internal dynamics contribute to the sustainment of the boundary over time. It also serves as the site of interaction between the autopoietic system and its environment. There is no requirement that the boundary be composed of any particular material.[20] Its importance lies in its dual role of differentiating the unified organism from its environment and (partially) defining the character of organism–environment interaction.

All of this serves as a response to the subjectivist challenge to agent self-determination raised in the previous chapter. That challenge depended on the idea that any stable dynamic system can be carved arbitrarily into parts, and that systems so described and other living systems differ only in their relative degree of complexity or interest for the observers doing the carving. The autopoietic approach to life demonstrates that living systems distinguish *themselves* as autonomous unities. Their organizational properties are dedicated to maintaining their operational closure and 'actively demarcating the boundary between 'self' and 'other'.[21] To say that the demarcation is active is to say that it requires energetic investment; the autopoietic system continually siphons order from its environment and channels it into the generation and maintenance of (1) an organizationally closed network of far-from-equilibrium processes and (2) a boundary that further distinguishes its autonomous dynamics from the environment and regulates material exchange between them.

Crucially, the autopoietic system is concerned with maintaining its organization. Maturana and Varela clearly distinguish the organization of a system from its structure. The organization of a system refers to the relations that define it as a member of a class of entities (this is the sense in which I have been using the term throughout this book). By contrast, the structure of a system refers to its actual material configuration. A given system may maintain a stable organization over a period of structural change, and a single organization may be multiply realized in various structures. Autopoiesis refers to the organization of the living – a set of relational properties that are realized in all living systems. When a living system is no longer able to generate its own components or maintain an operational boundary distinguishing its inner states from its environment – that is, its organization ceases to be autopoietic – it ceases to be alive. This is not a convenient 'as-if' description of the system – as features of the autopoietic system's organization, these activities are conditions on its very existence. The system's self-differentiation unfolds independently of whether some outside observer is present to interpret it as a unity.

From order to value

Thus we have a partial answer to the problem of self-determination. It offers an account of how living systems distinguish themselves from their surroundings.

However, it does not yet explain how *agents* distinguish themselves as such. Even if we restrict our focus to the living, it is unlikely that all living things are agents. Moreover, multicellular organisms are living things that are themselves made of other living things. You are a single agent despite being made of *trillions* of living things. The present account has not offered a principled means of distinguishing the agent. One is forthcoming, but it will require an answer to the problem of value.

Although organizational closure is not sufficient for life, it is a crucial feature of living systems: their recursive organizations are dedicated to their own maintenance and self-production. Jonas characterizes the living agent both in terms of its power to self-determine and its need. Its need arises from its organizational fragility; it assumes 'a position of hazardous independence from the very matter which is yet indispensable to its being'. Living systems work to distinguish themselves from their surroundings (a phenomenon demystified by appeal to the autopoietic organization), but ultimately 'life carries death in itself, not in spite of, but because of, its being life'.[22] This calls attention only to the bleak fact that all living systems inevitably disintegrate, but also the more urgent point that they will quickly do so in the absence of energetic input. The living organism is a process that is both wholly concerned with distinguishing itself from its environment and utterly dependent upon it:

> [I]ntrinsically qualified by the threat of its negative [life] must affirm itself, and existence affirmed is existence as a concern. So constitutive for life is the possibility of not-being that its very being is essentially a hovering over this abyss . . . being itself has become a constant possibility rather than a given state.[23]

The precarious character of life entails a primitive form of self-concern. This is due to the fact that living systems are energetically open, far-from-equilibrium systems, and thus their persistence depends on their continually drawing energy from their surroundings. Paradoxically, the living system cannot differentiate itself from its environment without engaging with it. Jonas claims that these conflicting demands are inseparable from the metabolic organization of the organism:

> This double aspect [of freedom and necessity] shows in the terms of metabolism itself: denoting, on the side of freedom, a capacity of organic form, namely to change its matter, metabolism denotes equally the irremissible necessity for it to do so. Its 'can' is a 'must', since its execution is identical with its being.[24]

This is the earliest and most fundamental form of purpose. It is important to note that the mere threat of dissolution is not the source of this need. If I decide to destroy my watch the moment it fails to keep time, I have not imbued the watch with newfound purpose. At best I have raised the stakes for the watch's fulfillment of an externally imposed function. By contrast, the function of autopoiesis *just is* the sustainment of autopoiesis. This circularity, captured in the notion of organizational closure, is the source of intrinsic normativity.

Evan Thompson elaborates this picture, drawing from Kant's discussion of organic nature in the *Critique of Judgment*. Kant believed that the existence of material objects could be derived from (and thus explained by appeal to) the application of Newtonian mechanics to physical matter. However, Kant also believed that the existence of living systems could not be derived a priori from the laws of physics because the existence of any organism is contingent relative to those laws. To explain the existence of an organism, Kant argued, one must appeal to purposes, in much the same way that we must appeal to extrinsic (observer- or creator-generated) purposes when explaining artifacts. However, unlike William Paley who took this fact to support idea that humans are a kind of divine artifact, Kant argued that the purposiveness of the organism was both intrinsic and 'natural'.[25]

Kant provides two criteria for natural purposiveness. The first is that 'the possibility of its parts [. . .] must depend on their relation to the whole', and the second is that 'the parts of the thing combine into the unity of a whole because they are reciprocally cause and effect of their form'. He restates this second condition in terms of self-organization shortly thereafter:

> Only if a product meets [these conditions], and only because of this, will it be both an *organized* and a *self-organizing* being, which therefore can be called a natural purpose.[26]

Kant believed that a full explanation of organismic activity requires an appeal to natural purposiveness. It should be noted that Kant did not view this normativity as *constitutive* of the organism, but rather as a regulative idea for an observer: to understand an organism, we are forced to view it as if it were a natural purpose. However, Andreas Weber and Francisco Varela have argued that Kant did not believe that talk of natural purposes was merely heuristic, which suggests that he realized that something intrinsic to the organism necessitated such judgments.[27] Thompson argues that Kant failed to elaborate this view only because he lacked the conceptual and mathematical tools of modern complexity theory.[28] Reflection on the process of self-production helps us understand why the living system's operations necessitate a teleological perspective. It is more than an 'eddy of order' – it is a wellspring of *value*. In the next section I show how this fundamental norm of self-production can be elaborated into a richer account of meaning construction, which in turn can be used to distinguish primitive agency from other complex processes.

Domains of interaction and enacted meaning

The viability of a system cannot be determined in abstraction from that system's environment. The most such an analysis could yield would be whether the system returns to certain stable configurations or not.[29] But unqualified structural stability cannot be the aim of the living system, as this would entail that the ultimate telos of any living system is death. No attribute is essentially advantageous: memory would be a burden in genuinely chaotic worlds; specialization of organ function is only good because we do not live in constant danger of being crushed or impaled (an environment that would favor homogenous systems).[30] This illustrates the essentially embedded character of the self-organizing system. Any evaluation of a system must specify the world it inhabits.

All self-organized behavior unfolds in an environment that will at any given time be conducive, neutral, or antagonistic to that unfolding. Here the value of a boundary in the autopoietic organization is obvious: it admits organization-sustaining elements and excludes harmful elements. It also serves as the sole site of interaction between the internal dynamics of the system and the environment. Thus in establishing its closure, the autopoietic system defines its environment in two ways, first by distinguishing its internal processes from the external world and the determining the environmental features to which it will be sensitive.

To see the complexity of this interaction, consider the relatively simple case of a glider in Conway's Game of Life. The Game of Life takes place on a two-dimensional grid or lattice. Each cell on the grid is either alive or dead, and as the game begins the state of each cell is determined by its proximity to other cells on the lattice at each time-step. The rules are simple: A dead cell with exactly three live neighbors becomes alive. A living cell with four or more living neighbors dies of overcrowding, and a living cell with fewer than two live neighbors dies (of loneliness, one supposes). A living cell with either two or three live neighbors survives to the next generation. These rules are sufficient to produce remarkably complex patterns of activity.

The glider is one very simple pattern that routinely forms in such systems. A glider is a pattern of five living cells that proceeds through a sequence of four transformations, alternating horizontal and vertical movements (see Figure 5.1). The result is a coherent pattern of activity that travels diagonally through its world. Randall Beer has used the Game of Life as a toy example to illustrate a number of core principles of autopoiesis, but for the present purposes one is essential, namely that of a *domain of interactions.*

Maturana and Varela define the domain of interactions for an organism as 'the set of all interactions into which an entity can enter'.[31] This domain can be divided into domains of nondestructive interactions (perturbations to the system that induce state changes but do not destroy its organization) and destructive interactions (which do destroy its organization). Because these domains are specified relative to system organization, it follows that different kinds of systems will construct different domains of interaction with their environment.

Moreover, the domains of destructive and nondestructive interaction for a single system are functions of its evolving internal state. As a 'structure of process' like the glider cycles through four states, the same stimulus may perturb it at one time and destroy it at another (see Figure 5.2). A nondestructive interaction may also have lasting effects on the trajectory of the glider, changing its phase and position and thereby priming it to respond differently to future interactions. This illustrates

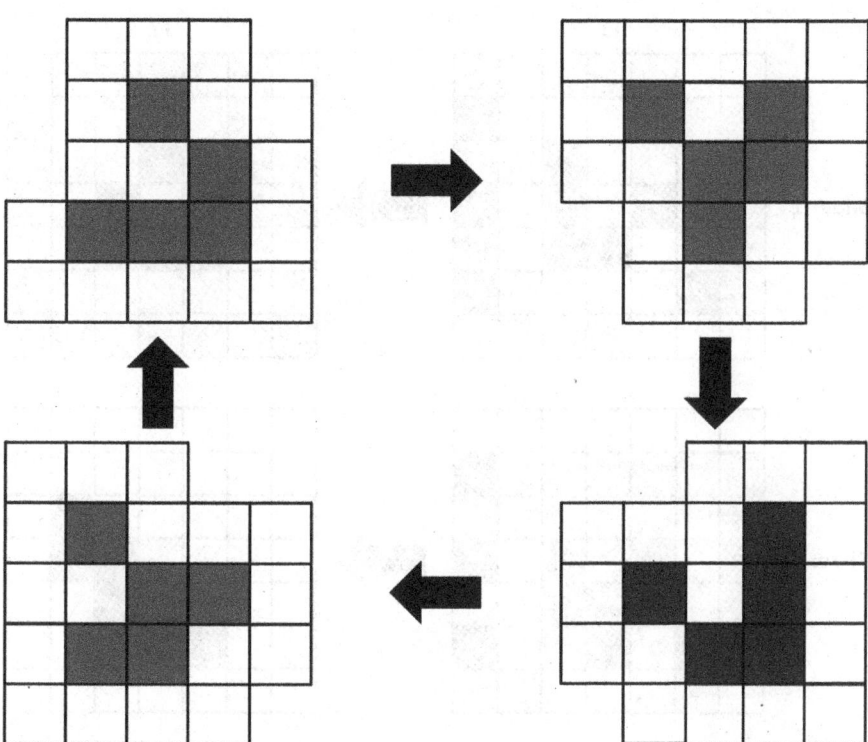

Figure 5.1 The characteristic four cycles of a glider in Conway's Game of Life.

the more general point that history of the self-organizing system can influence its internal state, which in turn affects its interaction with the world. Beer describes this feature as 'state-dependent differential sensitivity to perturbation'.[32] We might say that the structure of the system determines the set of possible state transitions that are available to the environment at any given time. The environment 'triggers' one of these transitions, which in turn causes a change in system state, potentially effecting some new change in the environment, and so on.

This reciprocal process of 'structural coupling' is continuous – at no time is a unity *not* interacting with its environment in this way. Thus the persistence of a system requires a degree of adaptive success, or 'good fit', with its environment. Crucially, survival for the glider is an all-or-nothing affair. It begins to dissolve the moment it has been perturbed out of its limit cycle. The system is either a good fit for its environment or it is dead. This is a feature of autopoietic systems in general, and indeed a problematic one to be addressed in the next chapter.

In a more detailed analysis Beer identifies 224 possible perturbations for the glider, which can be grouped into six categories on the basis of what effect they have on the glider.[33] These 'macroperturbations' determine the environmental impacts to which the glider is sensitive in virtue of its organization. Beer notes

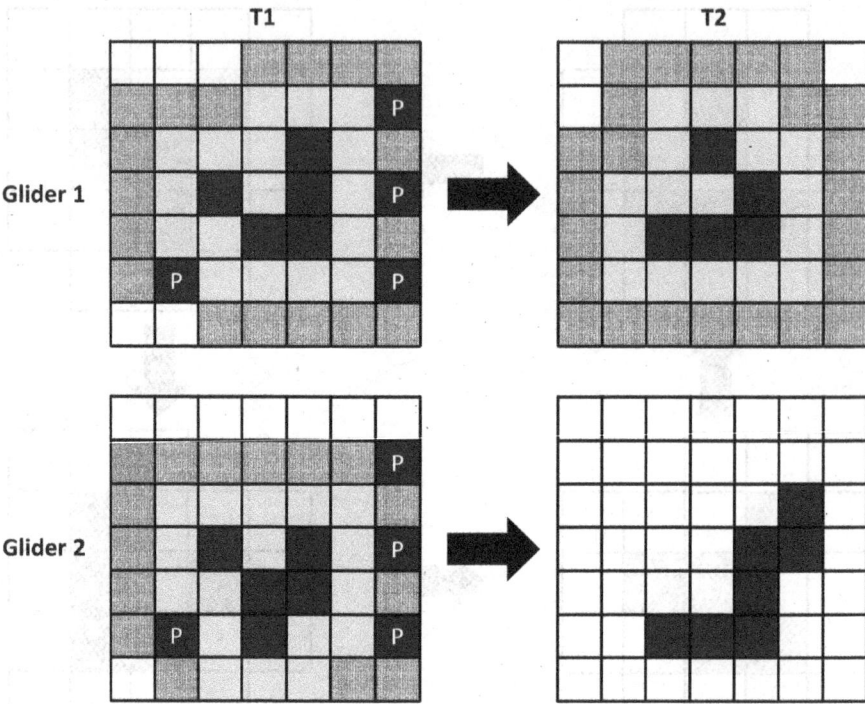

Figure 5.2 An illustration of state-dependent differential sensitivity to perturbation. The same perturbation (P) induces different changes in the glider depending on the glider's state at the moment of contact. This perturbation is nondestructive for Glider 1 but destructive for Glider 2.

that the majority of possible perturbations are fundamentally unknowable to the glider, falling well outside its viability constraint. The glider, in determining a domain of interactions with its environment, possesses a proto-perspective: certain features are harmful to it and others are benign.

This is analogous to what the physiologist Jakob Von Uexküll called the *Umwelt*: the subjective world within which the organism acts as agent. This world is defined in terms of the set of stimuli to which the organism is sensitive; the *Umwelten* of a human, a jellyfish, and a paramecium will be quite different. The analogy is not perfect, as Von Uexküll defined the Umwelt in terms of a set of receptor *and* effector cues.[34] Beer's analysis of the glider demonstrates how a detailed specification of an organism's subjective world can be constructed (at least in principle) on the basis of its structural coupling with its environment.

Importantly, the organization of the glider determines at each time step which features of its environment affect it and how. These categories of perturbation are not plain features of the Game of Life environment, but are jointly *enacted* by the glider and its environment. This evokes Maurice Merleau-Ponty's famous description of 'a keyboard which moves itself in such a way as to offer such or such of its keys to the in itself monotonous action of an external hammer'.[35] Value is not an objective feature of the environment (as it might be if the structure of the musical piece were found in the environment for the passive keyboard to receive), but is rather constructed by the system as it navigates its environment over time. Perception and action are 'never merely about the innocent extraction of information as if this was already present to a fully realized agent'.[36] Rather, the system and environment jointly enact meaning through their coupling.

This background has informed several detailed analyses of the behavioral dynamics of complex systems. The domain of nondestructive interactions for a system can be represented graphically as a space of possible stable configurations and their relations. The domain of destructive perturbations for that system can be treated as a 'viability constraint' that limits that space. Representing these domains as a continuous manifold, the behavior of the system can then be represented as a trajectory through the viable region.[37] This effectively specifies a norm of survival for a system – insofar as it is to retain its organization, its state must remain within its viability constraints.

Thus there are three ways in which the closed organization of the autopoietic system determines its subjective world. First, in distinguishing itself as a unity, the living system specifies its environment, much as in drawing a figure we thereby specify its background. The spatial boundary of the autopoietic system contributes to this differentiation. Second, the organization of the system determines its domain of interactions with the environment. Third, the system's organization will partially determine what stimuli it will actually encounter.

Xabier Barandiaran and Matthew Egbert extend this picture further in their analysis of norm-establishing dynamics in a model bacterium. Their target system, essentially an autocatalytic model coupled to a gradient-climbing chemotactic system, operates autonomously in an environment containing varying concentrations of food. Their analysis depicts the dynamics of the agent–environment system in terms of viable, precarious, and terminal regions (Figure 5.3). Living systems in the viable region will maintain their organization, and systems in the precarious region may survive, but only if environmental conditions change. Systems in the terminal region will die no matter what. Barandiaran and Egbert then construct a *normative vector field* for the agent, which indicates precisely how much of the food source an agent in various points of the precarious region must consume to enter or remain in the viable region (Figure 5.4). Model agents can then be evaluated by how closely their actual behavior maps onto the field specified by their embedded organization. Barandiaran and Egbert offer an operational definition of 'normative action' as 'system-driven modulation of its coupling with the environment whose effect on the viability space positively correlates with the normative field'.[38] Failed behavior correlates negatively with the field. Again, as with the case of the glider, this normative landscape is a function both of the dynamics of the agent and the state of its environment.

These cases are intended to show how the organizations of complex systems determine domains of interactions with their environments. But the mere fact that

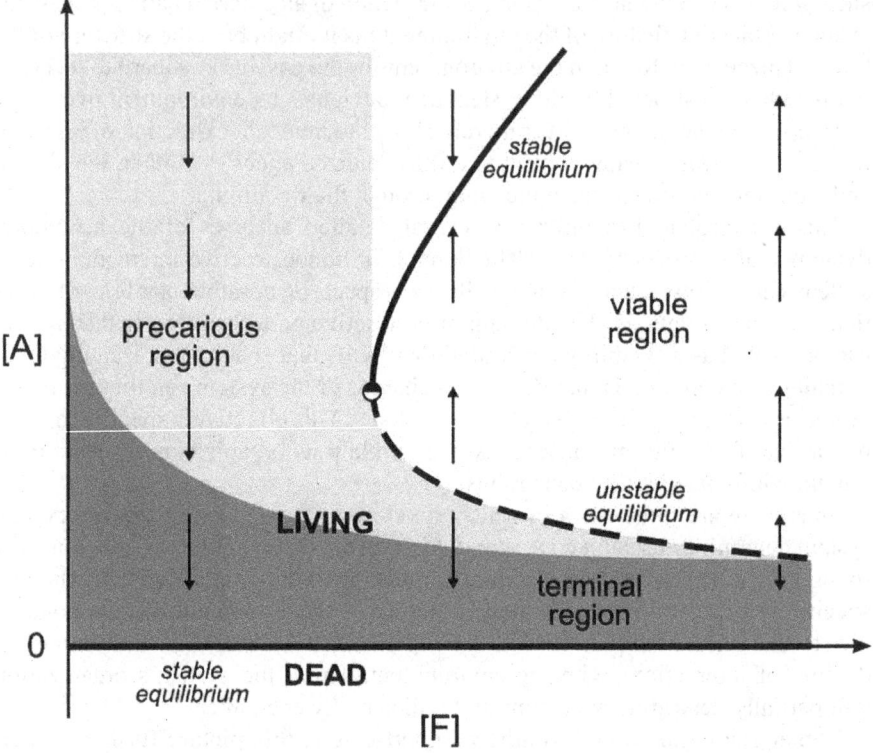

Figure 5.3 Viability space for an autonomous chemotactic agent.

such domains can be specified for a given system is not sufficient to show that the system actually has any such perspective. A similar analysis could be performed for a tornado. A considerably less interesting analysis could be performed for a rock. A glass will be perturbed out of its organization if it is struck with a hammer, but this fact does not matter to the glass, even as it shatters. Thus the fact that such regions can be specified for a system will not suffice to establish it as the sort of locus of value Jonas associated. Their normative force is *inherited* from the more basic norm of self-production.

Thus normativity arises in a more basic form than that addressed by traditional intellectualist approaches to mind and action. On such approaches, the normative character of an action is settled by the content of the motivating mental state. When I *try* or *intend* to raise my arm, the question of whether the activity is successful is determined by whether or not the trying or intending is satisfied (given certain ceteris paribus clauses designed to eliminate deviant forms of satisfaction). But here we see that the conditions of success and failure on environmental engagement are determined by the organization of the living system itself – the system either successfully copes with its environment or it ceases to exist. One could hardly imagine a stricter governing principle!

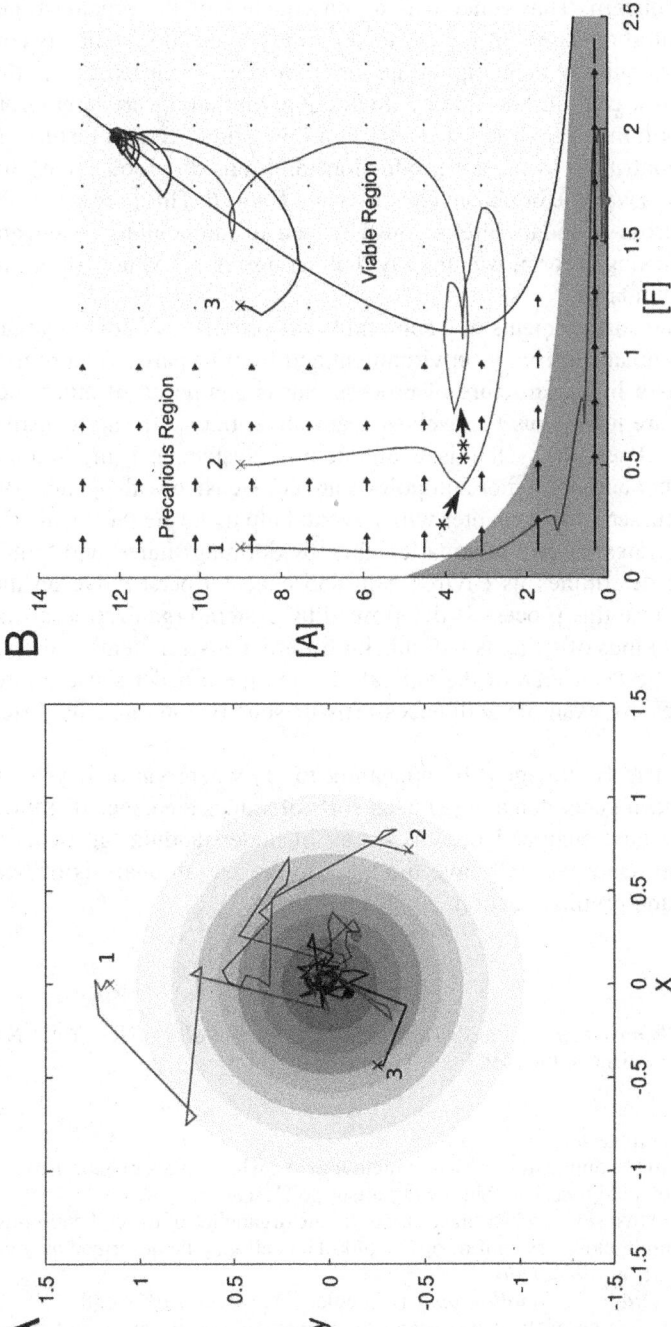

Figure 5.4 Barandiaran and Egbert's normative vector field for a stochastic gradient-climbing agent. (A) Three agents are started at varying distances from a food source. Their trajectories begin at the Xs and end at the closed circles. (B) The same trajectories projected within viability space. Agent 1's trajectory falls within the terminal region and it dies. Agents 2 and 3 remain within their viability constraint. The vectors indicate the minimum amount of constant increase in 'food' that is required to return to the viable region before dying. Agent 2's failures to correlate positively with the normative vector field are marked with * and **.

This perspective provides the means to explain the origin and role of natural purposiveness in a physical world. The living organization embodies the earliest form of self-concern. That concern is a consequence of the organizational closure and structural openness of the organism – the organism sustains its core invariant properties only by maintaining the far-from-equilibrium processes that maintain them, but it can only do that by interacting with an uncertain environment. Depending on the organism's structure at a given time, certain features of its environment contribute to its self-production (and thus are good for it) and others do not. In this way the organism constructs its *Umwelt*. This process results directly from the cross-boundary interactions between an autonomous system and its environment. In slogan form, we may say that the norms of living systems are enacted at their boundaries.

We now have the core elements of a solution to the problems of how to nonarbitrarily designate an agent from its environment and from its parts. The problem is that the agent is a living structure of process that is composed of other such structures, which are not agents themselves. I am alive, but so are my constituent cells. My central nervous system is an autonomous system, as is my immune system. Thus neither autonomy nor autopoiesis are coextensive with agency. But the principles discussed in this chapter will serve to help us locate the agent. The living system generates its own boundaries, thereby defining 'inner' and 'outer' processes. It both determines its environment and enacts a perspective on that environment. Because this process is determined by system organization, it follows that different kinds of systems will inhabit different environments and obey different norms. The *Umwelten* of the animal cell and the broader sensorimotor system it composes, for example, will necessarily be specified in radically different ways.

We may then identify the agent by appealing to (1) where various systemic boundaries have been generated and (2) what sorts of values are enacted at those boundaries. In the next chapter I present a way of understanding 'agent-level' value. With this in hand, we will have the tools to present an analysis of both primitive agency and primitive action.

Notes

1 H. Jonas, *The Phenomenon of Life: Toward a Philosophical Biology* (New York, NY: Harper & Row Publishers, Inc., 1966), p. 112.
2 Ibid., p. 3.
3 Ibid., p. 5.
4 Ibid., p. 79, my emphasis.
5 Jonas attributes this 'combustion theory of metabolism', which treats metabolism as a necessary but isolated process within the organism, to Descartes.
6 It should be noted that self-production ceases here; the organelles of the cell coproduce one another within a closed system of operations. The cell may be described as *autopoietic* and its parts as *allopoietic*.
7 D. Sands, *Introduction to Crystallography* (Mineola, NY: Dover, 1994), p. 81.
8 This is a characterization of the living organism, of course, not of what it is for anything to be an organism. There is nothing incoherent about the concept of a dead or

dormant organism. However, dead or dormant agents are incapable of anything like action or behavior.

9 E. Di Paolo, M. Rohde, and H. De Jaegher, 'Horizons for the Enactive Mind: Values, Social Interaction, and Play', in J. Stewart, O. Gapenne and E. Di Paolo (Eds.), *Enaction: Toward a New Paradigm in Cognitive Science* (Cambridge, MA: MIT Press, 2010), pp. 33–88, on p. 38.

10 See, for example, F. Varela, *Principles of Biological Autonomy* (Amsterdam: North-Holland, 1979).

11 For example, see E. Thompson, *Mind in Life: Biology, Phenomenology, and the Sciences of Mind* (Cambridge, MA: Belknap Press, 2007), p. 45.

12 Ibid., p. 65.

13 P.L. Luisi, 'Autopoiesis: A Review and a Reappraisal', *Naturwissenschaften* 90 (2003), pp. 49–59, on p. 52.

14 A. Juarrero, *Dynamics in Action: Intentional Behavior as a Complex System* (Cambridge, MA: MIT Press, 1999), p. 125.

15 K. Ruiz-Mirazo and A. Moreno, 'Basic Autonomy as a Fundamental Step in the Synthesis of Life', *Artificial Life* 10:3 (2004), pp. 235–259, on p. 240.

16 Hooker, 'Introduction to the Philosophy of Complex Systems', in C. Hooker (Ed.), *Handbook of the Philosophy of Science, Volume 10: Philosophy of Complex Systems* (Waltham, MA: Elsevier B.V., 2011), p. 35.

17 H.R. Maturana and F.J. Varela, *Autopoiesis and Cognition: The Realization of the Living, Boston Studies in the Philosophy of Science, Volume 42* (Dordrecht: D. Reidel, 1980), pp. 78–79.

18 Ibid., p. 98.

19 F.J. Varela, *El Fenomeno de la Vida* (Santiago, Chile: Domen Esayo, 2000). Cited in Luisi, 'Autopoiesis: A Review and a Reappraisal', p. 51.

20 The role of a spatial boundary in autopoiesis is the subject of some controversy, as it has been argued that the operational closure of an autopoietic system can extend beyond its physical boundary. See N. Virgo, M. Egbert, and T. Froese, 'The Role of the Spatial Boundary in Autopoiesis', *Advances in Artificial Life: Darwin Meets Von Neumann* 5777 (2011), pp. 240–247.

21 T. Froese and E. Di Paolo, 'The Enactive Approach: Theoretical Sketches from Cell to Society', *Pragmatics in Cognition* 19 (2011), pp. 1–36, on p. 7.

22 Jonas, *The Phenomenon of Life: Toward a Philosophical Biology*, p. 5.

23 Ibid., p. 4.

24 Ibid., p. 83.

25 Thompson, *Mind in Life: Biology, Phenomenology and the Sciences of Mind*, p. 133.

26 I. Kant, *Critique of Judgment* (Indianapolis, IN: Hackett Publishing Company, 1987), p. 253 (original emphasis).

27 A. Weber and F. Varela, 'Life after Kant: Natural Purposes and the Autopoietic Foundations of Biological Individuality', *Phenomenology and the Cognitive Sciences* 1:2 (2002), pp. 97–125.

28 Thompson, *Mind in Life: Biology, Phenomenology and the Sciences of Mind*, p. 138.

29 This is perhaps appropriate in certain analyses; for example, Anatol Rapoport defines a goal as 'some end state to which a system tends by virtue of its structural organization' in his mathematical analysis of general systems. Cf. A. Rapoport, 'Mathematical Aspects of General Systems Analysis', *General Systems* 11 (1966), pp. 3–11.

30 These examples are from W.R. Ashby, 'Principles of the Self-Organizing System', in H. Von Foerster and G.W. Zoph, Jr. (Eds.), *Principles of self-organization: Transactions of the University of Illinois Symposium* (London, UK: Pergamon Press, 1962), p. 265.

31 Maturana and Varela, *Autopoiesis and Cognition: The Realization of the Living*, p. 8.

32 R.D. Beer, 'Autopoiesis and Cognition in the Game of Life', *Artificial Life* 10:3 (2004), pp. 309–326, on p. 316.

33 R.D. Beer, 'The Cognitive Domain of a Glider in the Game of Life, *Artificial Life* 20:2 (2014), pp. 183–206, on p. 190.

34 J. Von Uexküll, 'A Stroll Through the Worlds of Animals and Men', in K. Lashley (Ed.), *Instinctive Behavior* (New York: International Universities Press, 1934), p. 12.

35 M. Merleau-Ponty, *The Structure of Behavior* (Boston, MA: Beacon Press, 1963), p. 13.

36 Di Paolo, Rohde and De Jaegher, *Enaction: Toward a New Paradigm in Cognitive Science*, p. 40.

37 Beer, 'Autopoiesis and Cognition in the Game of Life', pp. 320–322.

38 X.E. Barandiaran and M.D. Egbert, 'Norm-establishing and Norm-following in Autonomous Agency', *Artificial Life* 20:1 (2013), pp. 5–28, on p. 24.

6   From autopoiesis to agency

We are nearing an answer to the core problem of primitive agency, that of specifying the minimal conditions under which a system, considered as a whole agent, acts. I have proposed a general strategy for system discrimination that involves identifying the boundaries of a system through its domain of interactions with its environment and specifying the relevant norms governing those interactions. An account of primitive action requires the identification of a distinctly agential operational boundary and the specification of the norms governing interactions at that boundary. Action can then be defined in terms of those normative interactions. In this chapter I provide such an account. One additional refinement will be needed to proceed, however. Autopoiesis is necessary for agency, but it is not sufficient. We can see why by considering some of the recognized shortcomings of autopoiesis.

From autopoiesis to adaptivity

The first problem is that autopoiesis is an all-or-nothing affair. The sole norm governing the autopoietic organization demands the conservation of that very organization. But that norm is satisfied as long as the system remains within its viability constraints, even if its trajectory is rapidly approaching the boundary. There is no 'better' or 'worse' for a purely autopoietic system. It is not strictly bad for it to tumble over the edge of a cliff, for only the resulting impact is organizationally harmful.[1] There is no room for thriving, decline, threat, or improvement – the system either achieves a good fit with its environment or it disintegrates.[2]

The second problem is that although a norm-laden domain of interactions can be constructed for an embedded autopoietic system, autopoiesis does not require those norms to guide the system's behavior. It may establish the objective fact that a given system is about to exceed its viability constraint, but if the system itself has no access to this information, the fact that it really *ought* to modulate its internal state is of no use to it. Barandiaran and Egbert note this in their analysis of normative behavior: it is not enough that a system's behavior happens to correlate with its normative vector field – it must *modulate* its coupling with its environment in the service of those norms. This is the primordial version of the philosophical distinction between following a rule and merely behaving in accordance with it.

Finally, the self-concern engendered by the autopoietic organization can be satisfied in distinctly nonagential ways. Imagine three autopoietically viable organisms: Lucky, Rocky, and Ada:

> Lucky's existence is an ecological miracle. Its internal dynamics are entirely inflexible, but by sheer chance they exhibit a perfect fit with those of its environment. The slightest dissonant perturbation would send its dynamics careening out of the viable range, but no such event occurs. The world meets its various metabolic needs the moment they arise. Lucky satisfies its autopoietic norms almost by default. More bluntly, this organism never has to *do* anything to survive.
>
> Rocky exhibits equally inflexible behavior, but has the misfortune of existing in an inhospitable, chaotic environment that assaults it at every turn. Fortunately, Rocky is highly *robust* to perturbation, simply absorbing whatever its environment throws at it. Rocky's viability field is so broad that its dynamics will never exceed its constraints. This system maintains its invariant organization, thereby satisfying the autopoietic norm of survival, but it survives merely by absorbing impacts from its environment. This system never has to do anything either.
>
> Ada is neither lucky nor brutally robust. Its interactions with its environment vary in quality and occasionally drive it to the edge of its viability constraint. But Ada can register when this happens and modulate its dynamics accordingly, steering itself away from fatal trajectories. It may do this by actual movement in its environment or by adjusting its internal state. In either case, Ada exhibits norm-responsive state-change. This is to say that its behavior is *adaptive*.

Each of these organisms satisfies the demands of autopoiesis, but only Ada exhibits anything resembling agency. But Ada's defining characteristics – the ability to distinguish degrees of viability and modulate its structural coupling with its environment accordingly – are precisely the deficiencies in the pure autopoietic system identified in the first two problems. Ezequiel Di Paolo argues that adaptivity is necessary, not only for agency, but for the process of sense making in general; autopoiesis determines whether a feature of the world is good or bad for the organism, but this information only matters *to* the organism if it can inform the process of system-world coping.[3]

Di Paolo defines adaptivity as follows:

> A system's capacity, in some circumstances, to regulate its states and its relation to the environment with the result that, if the states are sufficiently close to the boundary of viability,
>
> 1   Tendencies are distinguished and acted upon depending on whether the states will approach or recede from the boundary and, as a consequence,
> 2   Tendencies of the first kind are moved closer to or transformed into tendencies of the second and so future states are prevented from reaching the boundary with an outward velocity.[4]

This requires interconnected processes of system-level monitoring and regulation. The adaptive agent does not merely succeed or fail, but also 'recognizes' when it is succeeding or failing and adjusts its behavior accordingly. Adaptivity does not require a high degree of cognitive sophistication, however: chemotactic bacteria can detect changes in attractant gradients, swimming in response to registered increases in nutrient concentrations and tumbling in response to decreases.[5] This requires an elaboration of basic autopoietic organization because the bacterium's behavior takes place within its viability constraint. A purely autopoietic system would have no reason to swim *up* a nutrient gradient at a given time if its needs were already being met.

Homeostatic adaptivity

Homeostasis, the maintenance of stability in the organism's internal environment, is a clear example of adaptive state change. In animals, this includes (but is by no means limited to) the regulation of 'defended levels' of blood glucose, core temperature, and water content of lymph fluid and blood. In some cases the operations of homeostasis are restricted to agent-internal processes but they commonly prompt animal behavior – feeding when hungry, drinking when thirsty, and so on. In such cases the behavior might be thought of as an extension of the organism's homeostasis.

Defended levels are often maintained through redundant effector mechanisms, and various considerations determine which mechanisms are used. For example, squirrel monkeys in a climate-controlled room maintained a favorable core temperature by repeatedly pulling a chain that changed the ambient temperature from 10 to 50 degrees Centigrade. When weights were added to the chain, however, the monkeys pulled the chain less often, relying more on the autonomic processes of shivering and increased metabolism to maintain their core temperature.[6] The increase in the energetic demand of a sensorimotor solution rendered it a less attractive solution to the adaptive problem of thermoregulation (a bad energetic *investment*, to foreshadow the next section), leaving the autonomic effector as the monkeys' preferred mechanism.

Homeostatic processes are concerned with defending a particular level of stability in the interior milieu of the organism, but organisms may also respond to external changes by changing that defended level through a process known as *rheostasis*. Rheostasis is evident in cases of seasonal hibernation, where body temperature and food intake are regulated at much lower levels than they are at other times of the year. It often serves to mediate conflicts between regulatory systems. For example, mammalian hibernators kept awake in laboratory conditions eat considerably less during the hibernation season, even when food is plentiful. This behavior seems maladaptive until one recognizes that hibernators periodically wake from hibernation to eliminate metabolic waste. If the animal's appetite remained at normal levels during its natural hibernation period, it would seek food upon awakening. More food means more waste to clear. Thus the animal would have to awaken more frequently from hibernation, at which point its appetite would drive it to search for food again. The scarcity of food in the winter months increases the energetic cost of foraging, further offsetting any benefit gained by hibernation. Programmed rheostasis solves this problem for the animal, effectively switching its appetite off during the hibernation period.[7]

This prescient adjustment of defended levels means that internal regulation requires more than environmental feedback. Homeostasis also makes use of regulatory feedforward networks. Nicholas Mrsosovsky notes that many animals drink water in anticipation of future demands. For example, rats drink most of their water during the first 30 seconds of feeding to preempt the dehydration caused by food intake. Pin-tailed sand grouses forage for water early in the day before the ambient temperature rises to a level that would require them to be hydrated. If these behaviors were only triggered by feedback, water foraging would not begin until a deficit had been registered.[8]

These cases demonstrate that not all forms of adaptive state change will count as action. The squirrel monkey may regulate its core temperature by either pulling a chain or shivering. Both are adaptive patterns of state change. But chain pulling is action, something the monkey does, and this is so even if its action is prompted by its internal thermoregulatory network. By contrast, shivering is an autonomic response to a drop in core body temperature. An immobile organism that was capable of dramatic adaptive responses in internal state could, through sheer mastery of its internal milieu, satisfy the demands of both autopoiesis and adaptivity. But its operations would not constitute anything like action – it would be massively adaptive but it would not *do* anything.

How do active patterns of adaptive state change differ from nonactive patterns? Here Di Paolo distinguishes the process of structural coupling from the regulation of that process, where regulation requires an 'asymmetry . . . in the domains of exchange between the unity and its medium'.[9] This asymmetry obtains 'when a process is established that is able to regulate [organism-environment coupling] so that in general the result is an improved condition of viability'.[10] Di Paolo then defines the agent as

> a self-constructed unity that engages the world by actively regulating its exchanges with it for adaptive purposes that are meant to serve its continued viability.[11]

I agree with much of this definition, but additional clarification is needed. The first worry is that the definition risks circularity: Agency is defined here as active regulation of organism–environment exchange, but the question of what *active* regulation amounts to is precisely the point at issue. We might defuse the circularity by understanding active regulation as asymmetric regulation, but the provided account of asymmetry defines it as a form of adaptive regulation, and it has already been established that adaptivity alone is insufficient for agency. Nor can the relevant sort of asymmetry be solely a matter of the agent's setting the terms of its ongoing interactions with the environment. If that were the case, then adaptive rheostatic processes like hibernation would count as action. What makes an adaptive modulation of structural coupling properly active?

Moving in the world

One plain response is that active adaptive behavior involves *moving around*. The monkey that pulls a chain to thermoregulate navigates a spatiotemporal environment to do so, whereas the shivering monkey does not (colloquially we would say that a shivering person was moving but not moving around). Movement through space is common to active adaptive system–environment interactions: arm raising, pursuit, avoidance, eating, chemotaxis, etc. Even behaviors like gliding, floating, or coasting involve the agent's allowing itself to be moved through space. Other adaptive processes that we would not classify as action – regulating blood pressure, hibernating, raising metabolic rate – are not concerned with spatial navigation. In acting our behavior is directed *at* the world rather than merely playing out *within* it.

This may seem obvious, but from the complex systems perspective it raises a curious problem. Reaching and shivering are both adaptive self-organized patterns of state change, driven by organismic need. Why should one pattern be privileged over the other as genuine action simply because it involves a particular sort of movement?

Because both behaviors are operationally closed processes, we can distinguish them by the elements involved in their closure. The boundaries of the shivering monkey's thermoregulative control loop are confined to its internal environment; its hypothalamus responds to monitored drops in core temperature by triggering vasoconstriction and the rapid, rhythmic contraction of skeletal muscle. It continues to monitor temperature through this process, terminating corrective measures once temperature stabilizes at the defended level. Its control loop stays within its internal milieu.

In this way this form of homeostasis is comparable to immune system function. The immune system is also operationally closed, modulating its internal state through a diffuse, decentralized cytokine network. Feedback through this network allows the system to identify pathogens and select appropriate responses from a range of effector types.[12] The immune system *does* have standards of proper function, derived from the living system's primary norm of self-production. It also contributes to the goals of the broader system of which it is a part, not only by protecting it from pathogens, but also by minimizing its energy expenditure. But its domain of interaction with the world outside its operational boundary is limited to interactions at that boundary with pathogens and information exchange with the autonomic nervous system. Its 'lived world', so to speak, does not involve movement through space. It is concerned with defending the interior from alien intruders, but it never strikes out into enemy territory.

By contrast, the monkey's chain pulling (or any thermoregulative *movement*, such as the homeostatic behaviors of ectothermic lizards), necessarily involves exploratory activity. This is a common feature of all active movement and requires a close link between perception and movement. As Kevin O'Reagan and Alva Noë argue, successful perception requires mastery of *sensorimotor contingencies*: the rules governing the relationship between sensory inputs and motor outputs.[13] In vision, this includes the rules governing contingencies as small as the retinal distortions induced by eye saccades. Successful hearing involves not only receiving sound waves, but also tracking the changes in temporal asynchrony and perceived amplitude of auditory signals induced by head movement. Embodied movement alters the perceptual field, which in turn requires adjustments to that movement, and so on. Stable behavior results from a kind of functional integrative competence involving continuous processes of monitoring and adjustment over its performance. Success in the form of stable, adaptive behavior is achieved when the requisite sensorimotor loops are closed and the global behavior achieves operational closure. From the systems perspective adopted here, this closure depends on the enslavement of local processes into coherent, adaptive wholes. Movement through space is a process of constructing a synergy of interdependent perceptual and motor components – a *sensorimotor structure of process.*

This brings us a step closer to the goal of identifying the agent. First, if the domain of interactions between an agent and its environment involves this sort of constructed movement, then the relevant systemic boundary for agency is the operational boundary of the living sensorimotor animal. It entails the existence of a distinctly agential domain of interactions and a unique set of corresponding norms. Reaching and shivering are subject to different norms of performance, even if they are performed in the service of the same broad norm of internal self-regulation. Whereas the failure to thermoregulate by shivering results either from malfunction or an unmanageable drop in core temperature, there are many ways to fail to regulate one's core temperature by pulling a chain (failing to grasp the chain, encountering a sudden obstruction, discovering the chain requires too much effort to pull, mechanical failure in the air conditioning system).

In some these cases the success or failure of the movement is not fully up to the agent. Even in such simple cases as arm raising, the agent operates within a medium that it can influence but never fully control. Successful movement always requires a degree of cooperation from the environment, and that cooperation is not always guaranteed. This affects the way that the sensorimotor system must interact with its environment to remain viable. In the next section I show how the unique demands of adaptive movement in space constitute the most basic forms of agential normativity.

Sensorimotor normativity and the organization of the agent

Jonas argues that the ur-norm of survival, although essential to all autonomous biological life, cannot make sense of sensorimotor animal life. The norms governing animal movement are related to those governing organic life generally. The difference lies in the way that the motile organism is situated relative to the materials it needs to perpetuate its existence:

> The great secret of animal life lies precisely in the gap which it is able to maintain between immediate concern and mediate satisfaction, i.e., in the loss of immediacy corresponding to the gain in scope.[14]

Jonas contrasts the animal organization with that of the plant, for which 'need passes of itself over into satisfaction by the steady operation of its metabolic dynamics'.[15] Nonmotile life either exploits the materials encountered at its immediate boundary or dies. By contrast, the organization of the motile animal has evolved to cope with a world that holds its life-sustaining materials at a spatiotemporal distance. In other words, its organization *dynamically presupposes*[16] the spatiotemporally extended nature of its environment – it is built to move. This has consequences for the domain of interactions between the agent and its environment and the corresponding values enacted at their shared boundary.

Jonas characterizes the shift from basic metabolic need to what he calls *appetite* as a gradual increase in the 'transcendence' of life beyond its 'point-identity'.[17] All life, depending upon material exchange with its environment to perpetuate its identity, faces 'forward as well as outward . . . beyond its own immediacy'.[18] This 'transcendence' amounts to dynamic presupposition along both spatial and temporal dimensions: the motile organism has evolved to depend upon an energy source that lies well beyond its physical boundary. Its domain of interactions with its environment reflects this fact. Phenomenologically speaking, the motile animal inhabits and acts within a spatiotemporally extended *Umwelt*. The scope of the nonmotile system's lived world, by contrast, is limited to the edge of its physical boundary.

Jonas notes that the spatiotemporal character of motile life entails a gap dividing 'action from its purpose',[19] specifically the metabolic purpose for which all action is ultimately performed. This gap requires the animal to perform 'intermediate' movements that contribute to metabolism only indirectly through their results. These movements draw upon the organism's surplus energy, 'an expenditure to be redeemed only by [its] eventual success'.[20] This unique feature of motile life places additional demands on the organism. Intermediate movement is neither better nor worse from the purely autopoietic perspective, but an organism with finite energy reserves must view it as such. Intermediate movement comes at an energetic cost, and success in an unpredictable environment is not assured. Animal movement essentially involves gambling with one's energy reserves in hopes that the environmental payoff will have been worth the risk.

This is reminiscent of Millikan's embedding requirement for functional behavior, discussed in Chapter 3. Millikan argued that functional behavior differs from other functional state changes in that it effects changes in the organism's environment in order to recoup some benefit from those changes. This is why, for example, a clam's slowing its activity in cold water does not count as functional behavior but a spider's pursuit behavior does: only the latter involves an energetic *investment*, rather than a mere expenditure.

Actions as spatiotemporal structures of process

We are now positioned to understand the asymmetric character of agent–environment interactions and the norms governing their operation. The 'active modulation' of structural coupling should not be taken to mean 'regulated by the agent', but rather 'requiring an asymmetric energetic commitment on the part of the agent'. This commitment goes well beyond merely putting energy out into the environment and hoping for the best, however. It requires investing it in the construction of intermediate movements, understood as morphologically coherent, operationally closed, sensorimotor processes. The sensorimotor agent must be able to evaluate its intermediate movements as more or less likely to achieve its metabolic ends and, in some cases, to modify or terminate movements that it determines to be bad investments over the course of their performance.

This can be elucidated by reflection on the outward manifestations of shivering. Shivering is an adaptive state change initiated as part of an organism-internal thermoregulatory feedback loop, and is decidedly not active. Of course, the oscillatory trembling characteristic of shivering does have some outward energetic output. With that in mind, consider the following case:

> An unfortunate dog has lost all nervous function, save for basic autonomic processes. It is entirely unaware of its surroundings. Among its problems is the fact that it lives in a space of widely varying and fluctuating temperatures. Thus, it needs to periodically move from place to place to keep its core temperature within an appropriate range. An engineer creates a machine to assist the dog in this task – essentially a motorized, wheeled platform with a sensor. The dog sits atop the platform. When its core temperature drops, it shivers. The vibrations caused by the shivering trigger the sensor, which initiates a random walk sequence, moving the dog to another location. Eventually the platform reaches a warmer area and the dog stops shivering, which in turn halts the machine.

Although the dog's shivering is adaptively successful because of an agent-to-environment flow of energy, the dog is not acting. The energetic byproduct of the shivering does work and ultimately serves the dog's adaptive needs, but the shivering itself remains a self-contained process, and is no more active than it would have been without the machine (the coupled dog–machine system, on the other hand, could, with some modification, be a plausible thermotactic cyborg!).

The proper sort of energetic commitment can be explained in terms of the self-organization of sensorimotor patterns of state change. Movements are synergies – morphologically coherent, operationally closed, dissipative structures of process. Whereas internal homeostasis involves channeling energy into existing processes, movement involves the investment of energy into the generation and maintenance of *new* dynamic structures. In reaching for the chain, the monkey – itself a self-organized system – sacrifices some of its order to construct and sustain a localized sensorimotor structure that functions to transfer energy to the environment in the form of physical work. Walking is a self-organized structure of process assembled and fueled by the organism for purposes of locomotion. The interaction is asymmetric because the environment does not initially match the energetic investment required to produce and maintain that structure – the payoff comes only if the (potentially long-term) environmental response to the movement is favorable. By contrast, the system-internal process of shivering immediately pays what it can back to the organism in terms of heat energy.[21]

In saying that agents construct their actions, I do not mean that actions are separable from their agents. That would open the door to the very deviance problems that motivated the shift away from the efficient-causal picture of action. The context-sensitive constraints underlying the self-organization of an agent's behavior occur within its boundaries. The power we have over our own movements is fundamentally different from the control we have over the movements of the objects around us. A movement is a new, operationally closed synergy assembled within the agent's operational boundary. It includes not only the limbs being moved, but the sensorimotor system as a whole. My arm's going up is merely the externally observable part of a larger organized process.

Thus a division in biological norms is warranted. Whereas metabolic norms apply specifically to the persistence of the organism – they are satisfied just when the self-producing system remains within viability constraints – agential norms apply to the investment of energy in the construction of adaptive sensorimotor synergies. They are satisfied when the organism gets a positive return on its energetic investment. They are not when the investment fails to pay off, leading to a decrease in overall system viability.

It is important to remember that although autopoiesis is insufficient for agency, it *is* necessary. The force of these agential norms is derived from the original norm of self-production. The agent's *Umwelt* enriches (but does not replace) the autopoietic perspective: stimuli are not merely 'good' or 'bad', but also 'worth it' or 'not worth it' from the perspective of an embodied sensorimotor agent with finite energy reserves, which must meet its needs by generating new order in an uncertain environment. This perspective is the hallmark of primitive agency.

Multicellular life and the demands of agency

Successful sensorimotor interaction requires that behavioral synergies be not merely constructed but *sustained* through the course of their performance. This does not mean that the agent must control every aspect of its movement, however. As demonstrated by Thelen and Smith's case of rhythmic infant kicking, structural features of the body and environment will play a role in the self-assembly of behavior. But this does not preclude the agent from generating sensorimotor structures that incorporate those features. Even ballistic movements that include an uncontrollable component (praying mantis striking, chameleon tongue projection, mantis shrimp spearing) are preceded by a sustained period of sensorimotor environmental coping. For example, the praying mantis gauges prey distance by integrating information from both eyes and consistently adopts attack or defense postures as a function of distance, adopting a striking posture when the distance drops to less than 100 mm, striking at prey between 15 and 30 mm, and striking defensively when objects are closer than 15 mm.[22] The strike itself is ballistic, but it is merely a component of a more complex integrated sensorimotor whole. Moreover, during the ballistic portion of the movement the organism must make continuous postural adjustments to compensate for its force. Similar things could be said for aimed defensive chemical responses such as those used by skunks and bombardier beetles.

The integrated nature of sensorimotor activity assures some real-time reactivity and flexibility. Mantises can transition from attack to defensive postures as their needs demand, attacking lions are able to adjust their pursuit in response to their prey's evasive maneuvering, and bacteria respond adaptively to changes in nutrient gradient by reversing the rotation of their flagella. Different kinds of systems will satisfy agential norms in different ways. Whereas bacterial movement is a sensorimotor process in that the bacterium modulates the direction of its flagellum's rotation on the basis of sensory input, the organization of the flagellar whipping pattern is settled by the material properties of the flagellum and the containing medium. There are no internal states in the bacterium devoted to overseeing the whipping, nor need there be.

By contrast, the construction of embodied human movement requires the participation of more elaborate feedback and feedforward systems distributed throughout the agent. This is due to the fact that human bodies have considerably more freedom of movement than bacterial bodies and thus require additional constraint to generate adaptively successful synergies. But the same norm of agency governs both systems. Both the human and the bacterium must, on pain of death, invest energy in the construction and sustainment of successful sensorimotor behavior in an uncertain environment. It simply happens that the human requires more machinery to be in place to make that investment possible.

The difference in the ways that unicellular and multicellular agents satisfy norms of agency can be seen in the difference in the number of processes that are subordinated to the goals of the whole organism, as seen in the increased differentiation of organ function in larger animals. This integration is not shared by all motile multicellulars, however – 'swarm' systems such as *V. carteri* and *M. xanthus* behave as functional collectives but their behaviors are essentially the vector sums of their constituent cells' individual behaviors. Nothing like global integration takes place.[23]

The requirement that action be a kind of sensorimotor construction does not return us to a centralized picture of cognition and action. Although it is true that in many cases genuine sensorimotor behavior requires processes that loop through the central nervous system, it is not necessary. To use an earlier example, octopus movements take multiple forms. Octopuses do not always rely on tentacle–eye coordination to grasp; the articulated movements involved in grasping are handled via sensorimotor networks in the peripheral nervous system. However, locomotion, which involves the selective activation of semiautonomous tentacle networks, does require the integration of visual and motor information in the central nervous system. Both behaviors are actions, however – they are operationally closed, sensorimotor constructions fueled by and subordinated to the global organization of the octopus as a whole. Both obey agential norms of investment – they require the investment of energy in the construction of intermediate movement in an uncertain environment and are successful when that investment pays off.

The idea that behavior is a constructed sensorimotor process helps us handle a surprisingly difficult borderline case of multicellular agency. Jonas contrasts motile life with the rooted life of plants, and we might be intuitively loath to describe plants as agents. But a surge in recent work on adaptive plant behavior has suggested that this question is not so easily answered. Plant behavior ranges well beyond the stock example of phototropism. Plants compensate for their immobility with remarkable phenotypic plasticity, modifying their growth patterns in adaptively advantageous ways. Many plants employ elaborate chemical signaling mechanisms to communicating with one over long distances. For example, African acacia trees have evolved the ability to alert one another to the presence of predatory kudu antelope. Upon being nibbled, the acacia increases its tannin production to near-lethal levels. More remarkably, the acacia releases ethylene into the air that causes downwind acacias to increase their tannin production within 5 to 10 minutes.[24] This is an adaptively successful form of signaling, and indeed the states involved, may count as having intentional content on some teleofunctional accounts.[25]

Plants also exhibit a range of sophisticated foraging behaviors both above and below ground. As with animal movement, plant foraging is an exercise in energy management. Plants adjust their growth patterns to maximize nutrient uptake at minimal energetic cost. Exploratory roots require resources that could otherwise be used for leaf or seed production, but if they tap into nutrient-rich soil, the whole plant benefits. Plants also adjust their growth to adapt to different soil volumes; if two plants are given the same water and mineral resources but placed in pots of different sizes, the plant in the larger pot will show increased growth.[26] If an individual plant is grown with its roots divided between two boxes, one containing competitor plants and one without any competition, the plant will devote its energy to growing roots in the vacant box, allowing the roots in the other box to wither.[27] Plant behavior, rather than being purely environment driven, is a slow but continuous process of resource investment and space management in the service of whole-system goals.

This suggests that plants deserve serious consideration as agents. Their designation hangs on two questions. The first is whether oriented growth is an acceptable substitute for movement. Because the processes of control associated with animal movement are not present in plants, exploratory root formation has a different structure than animal movement. Argyris Arnellos and Alvaro Moreno argue that sensorimotor agency requires the participation of both constitutive (self-producing) and interactive (behavioral) processes. The plant's interactions, with the exception of signaling, *just are* its constitutive processes, and consequently operate on slow developmental, rather than quick sensorimotor, timescales.[28] All of this disqualifies plants from inhabiting *our* shared *Umwelt*. But the broad similarities between plants and other agents might suffice to qualify them as agents of a very different sort that adaptively grow into their spaces instead of moving through them. If this is so, we might think of them as constructing structures of process in their environments as well, albeit more permanent structures than arm raising.

The second question involves the degree to which plants can be said to integrate information. Root and shoot growth in plants is largely sectorial and competitive. Extensions of the plant that find fertile soil or better exposure are rewarded with more resources. This raises the worry that plants are not integrated functional wholes but competing collectives.[29] Scientific work on this topic is ongoing, and I will not settle the debate here, but it is noteworthy that the primary arguments for the agency of plant life emphasize the features that I have argued are central to primitive action – movement in space, integration of sensory and growth processes, and energy management. This suggests that we are on the right track.

Primitive agency

I can now offer a definition of primitive agency. The primitive agent is an organizationally closed, adaptive sensorimotor system. Its domain of interactions with its environment includes interactions associated with the performance of intermediate movement in the world to acquire distal resources. Those movements require the investment of energy in the construction and sustainment of sensorimotor synergies that modulate the agent's coupling with its environment in adaptive ways. Those interactions are governed by norms of energy investment, which acquire their force from the fundamental biological norm of self-production.

With this in hand we can return to the criteria for action and agency formulated in Chapter 1.

A1. Actions are produced by agents, not by mere parts of agents or external forces

Agents are self-organizing systems that both comprise and compose other self-organizing systems. The matter of determining what system is behaving at a given time can be settled by identifying an appropriate domain of interactions and the norms governing that domain. Here appropriate domain of interactions holds between an organizationally closed sensorimotor system and its environment. The movements of this system are characterized in terms of the interactions at that system's boundary. The *actions* of this system are those sensorimotor-level interactions that are governed by norms of energetic investment, that is, events where the system invests energy in the generation and sustainment of perceptuomotor synergies in order to obtain organization-sustaining elements from its environment. This account distinguishes digesting from feeding, dancing from spasming, knee-jerk reflex behavior from kicking, and sweating from turning on the fan. It preserves the appropriate distinction between part-attributed behavior and whole-attributed behavior without adverting to the propositional attitudes.

A2. Actions are coordinated behaviors

Both agents and their actions are self-organized systems. Action involves the enslavement of local parts into adaptive sensorimotor synergies. This demand necessitates a functional competence in managing sensorimotor contingencies, which in turn require energizing, structuring, and sustaining these synergetic movements in real time.

A3. We can intervene to stop our basic actions

The sensorimotor competence required for savvy energetic investment requires the ability to restructure sustained movements in real time in response to changes in internal or external state. At times that may involve denying a maladaptive sensorimotor structure of energy, terminating the movement. This applies to activities at varying scales; I stop reaching for the stove when I realize it is hot; the lion stops pursuing the wildebeest when it escapes into the herd; the squirrel monkey stops pulling the chain when it becomes prohibitively difficult to pull. In these cases metabolic concerns motivate the choice of one course of action over another. In other cases, to be discussed in the conclusion of this book, other forms of stability may warrant the production or dissolution of movement.

A4. We have nonobservational access to our actions

The epistemic features of action, A4 and A5, may not obtain in the most basic systems. It is unclear that we ought to view a housefly as possessing self-knowledge or as being *surprised* by anything. But primordial versions of these capacities are present in primitive agents. For example, the requirements for the sustainment of sensorimotor behavior structures guarantee a primitive form of agent's knowledge. The agent's domain of interaction is partially constructed by that agent, the structure of which exhibits a high degree of fit with its environment. The movement is directed at the world but constructed within the operational boundaries of the agent. The agent, qua organism, is not distinct from its behavior, but rather supervenes on the temporally extended sum of its state changes. Thus it is related to the construction of its movement in a way that is distinct from its relations to any other process. The sustainment of an agent's movement requires a functional competence with the relevant sensorimotor contingencies within its domain if of interactions. In human agents this requires the operation of both feedforward and feedback systems, but other agents may satisfy the requirement in more basic ways. Either way, the agent necessarily tracks its own behavior in sustaining it in a distinctly non-observational way.

A5. In acting, we are unsurprised to observe that we have acted and uniquely surprised when our actions are thwarted

Thwarted action amounts to a case of bad fit between agent and environment. Here the notion of dynamic presupposition is helpful: dynamic coupling involves modulating one's own dynamics in a way that presupposes a compliant world. Alas, sometimes the world does not comply. This 'unexpected' dissonance between the states of the most primitive agents and those of the environment can be seen as the precursor of more sophisticated cases of surprise. Genuine surprise may result from the meta-awareness of such coupling failures.

A6. Action has success and failure conditions

Primitive agency is essentially normative. Indeed, the organizational closure associated with life engenders the first form of nonderived normativity. Actions are governed by norms of investment which inherit their force from the ur-norm of organizational conservation. The self-producing, adaptive system is the origin of subjective value and of agency generally.

Notes

1 This example is taken from E. Di Paolo, 'Autopoiesis, Adaptivity, Teleology, Agency', *Phenomenology and the Cognitive Sciences* 4 (2005), pp. 429–452, on p. 436.
2 Thus Barandiaran and Egbert's 'normative vector field', which prescribes corrective behavior for the system in various regions of its viability space, is properly understood as an elaboration of autopoiesis.
3 Di Paolo, 'Autopoiesis, Adaptivity, Teleology, Agency', pp. 438–439.
4 Ibid., p. 438.
5 R.M. Macnab and D.E. Koshland, Jr., 'The Gradient-Sensing Mechanism in Bacterial Chemotaxis', *Proceedings of the Natural Academy of Sciences* 69:9 (1972), pp. 2509–2512.
6 E.R. Adair and B.A. Wright, 'Behavioral Thermoregulation in the Squirrel Monkey When Response Effort Is Varied', *The Journal of Comparative Physiology and Psychology* 90:2 (1976), pp. 179–184.
7 N. Mrosovsky, *Rheostasis: The Physiology of Change* (Oxford, UK: Oxford University Press, 1990), pp. 36–38.
8 Ibid., p. 5.
9 Di Paolo, 'Autopoiesis, Adaptivity, Teleology, Agency', p. 443.
10 Ibid.
11 Ibid.
12 J. Brownlee, 'Autonomous Distributed Control in the Immune System Using Diffuse Feedback', *CIS Technical Report,* 070329A (March 2007), at http://citeseerx.ist.psu.edu/viewdoc/download?doi=10.1.1.71.1657&rep=rep1&type=pdf [accessed 15 May 2015].
13 J.K. O'Regan and A. Noe, 'A Sensorimotor Account of Vision and Visual Consciousness', *Behavioral and Brain Sciences* 24 (2001), pp. 939–1031, on pp. 940–941.
14 H. Jonas, *The Phenomenon of Life: Toward a Philosophical Biology* (New York, NY: Harper & Row Publishers, Inc., 1966), p. 102.
15 Ibid.
16 The term is borrowed from Mark Bickhard: 'That is, it is derived from the norm of contributing to the maintenance of the conditions for the far from equilibrium continued existence of the system . . . More generally, a process dynamically presupposes whatever those conditions are, *internal to the system or external to the system,* that support its being functional for the system.' M.H. Bickhard, 'The Dynamic Emergence of Representation', in H. Clapin, P. Staines and P. Slezak (Eds.), *Representation in Mind: New Approaches to Mental Representation* (Oxford, UK: Elsevier, 2004), pp. 71–90, on p. 77 (original emphasis).
17 Jonas, *The Phenomenon of Life: Toward a Philosophical Biology,* p. 85.
18 Ibid.

19 Ibid., p. 104.
20 Ibid.
21 This is closely related to a distinction made by Alvaro Moreno et al. between constitutive and interactive processes. Whereas constitutive processes are concerned with self-organization, the latter are involved with behavior generated in order to modify detected environmental conditions so as to preserve or improve viability. The agent is composed of both; it is an organizationally closed system that generates operationally closed perceptuomotor systems to serve its organizational needs. A. Moreno, A. Etxeberria, and J. Umerez, 'The Autonomy of Biological Individuals and Artificial Models', *BioSystems* 91 (2008), pp. 309–319.
22 S. Rossel, 'Binocular Spatial Localization in the Praying Mantis', *The Journal of Experimental Biology* 120 (1986), pp. 265–281, on pp. 274–275.
23 A. Anellos and A. Moreno, 'Multicellular Agency: An Organizational View', *Biology and Philosophy* 30:3 (2015), pp. 333–357.
24 S. Hughes, 'Antelope Activate the Acacia's Alarm System', *New Scientist* 127 (1990), p. 19.
25 Cf. R. Millikan, 'Biosemantics', *The Journal of Philosophy* 86 (1989), pp. 281–297.
26 K.D.M. McConnaughay and F.A. Bazzaz, 'Is Physical Space a Soil Resource?' *Ecology* 72 (1991), pp. 94–103.
27 M. Gersani, Z. Abramsky, and O. Falik, 'Density Dependent Habitat Selection in Plants', *Evolutionary Ecology* 12 (1998), pp. 223–234.
28 A. Anellos and A. Moreno, 'Multicellular Agency: An Organizational View', *Biology and Philosophy* 30:3 (2015), pp. 333–357.
29 T. Sachs, A. Novoplansky and D. Cohen, 'Plants as Competing Populations of Redundant Organs', *Plant, Cell and Environment* 16 (1993), pp. 765–770, on p. 765.

Conclusion
Beyond primitive agency

In this book I have attempted to answer Frankfurt's challenge for the philosophy of action and provide a broad, unificatory account of agency that is common to much of the biological world. A truly complete account of action would demonstrate how the raw materials of primitive agency are refined into the full-blooded intentional forms that occupy so much attention in the philosophy of action. That project is beyond the scope of a single book, and I will not attempt it here. But one small step in this direction should be made.

A reader who has made it this far will surely have objected at some point that not all of our actions are aimed at satisfying our metabolic needs, even indirectly. Hunger strikes and suicides are actions, after all, and only rarely does arm raising have any bearing on our survival, to say nothing of mindless habits like finger drumming and idle pacing. How are these to be understood on the account I have proposed?

The answer lies in understanding what would constitute a proper return on energetic investment. As Jonas notes, 'the survival standard' is inadequate as a norm for all living movement. His explanation for this, however, is cryptic:

> The feeling animal strives to preserve itself as a feeling, not just a metabolizing entity, i.e., it strives to continue the very activity of feeling: the perceiving animal strives to preserve itself as a perceiving entity – and so on.[1]

This remark can be elucidated by looking at the history of organismic development. Alvaro Moreno and A. Lasa argue that the demands of active sensorimotor coupling placed additional selection pressures on organisms as they increased in size. Specifically, the demands of coordinating complex real-time adaptive behavior in larger animals require the evolution of dedicated information-processing subsystems that are increasingly decoupled from basic metabolic processes.[2] Whereas the earliest forms of adaptivity are nearly indistinguishable from metabolic processes, larger multicellular animals require fast and efficient intercell communication networks to generate adaptive whole-system behavior.

The evolution of nervous systems in eumetazoans marked a substantial developmental leap forward. Neurons are metabolically efficient and highly plastic. As a result, even basic nervous systems tend to display recurrent activity. That recurrence results in the establishment of new operationally closed systems, effectively decoupling their informational patterns from the underlying rules of metabolism.[3] Continued growth led to continued division and specialization of the nervous system, to include the separation of the autonomic and limbic systems in vertebrates. These semiautonomous systems are subject to their own stability conditions. However, the nervous system and the metabolic network of the organism are interdependent: the metabolic components of the organism rely upon the nervous system to accomplish its needs, whereas the metabolic organization in turn supports the maintenance, construction, and functioning of the nervous system.

Thus larger vertebrates share similarities with basic autopoietic systems but are also importantly different in a way that explains the environment-independence of sophisticated sensorimotor behavior. The requirements for stability of a decoupled neural network may be quite abstract: the decoupled system may be concerned more with the self-maintenance of habits, *rationally* coherent behavior, psychodynamic stability, and so on.[4] Social, language-using animals place further demands on these networks, which have come to exhibit patterns of neural activity that most of us describe with the language of propositional attitudes. At this level of sophistication, we find *thinking things*. But this is not a return to Cartesianism: these networks obey norms governing their own dynamics but ultimately are both composed of and serve living systems. The living organization remains the source of original meaning and purpose, and the norms governing these decoupled networks inherit their force from its structure.

As biological systems become more complex and their subsystems become progressively autonomous from their basic biology, their needs may take on a different character. Many of our 'mindless' idle activities, for example, may be simple exploratory behaviors performed for the purpose of increasing the probability of encountering novel stimuli. Others may have the function of discharging excess energy to avoid overloading other networks. Still others, such as rhythmic finger drumming, may function to maintain fine motor network function. This is a way of elucidating Jonas's claim that the feeling organism has a need to preserve itself as such. But in all such cases the fundamental norms of agency hold – all such behaviors are constructed sensorimotor synergies, the success of which are settled by whether norms of energy investment are satisfied. The varied stability requirements of these decoupled networks simply change what would count as an appropriate return on investment. In the most basic forms of action, the return on investment is an immediate contribution to the processes of self-production that constitute the system. In more sophisticated agents comprised of multiple semiautonomous networks, the return may be stability of another kind such as guaranteeing rational coherence of one's attitudes about the world.

The framework I have presented offers a way of distinguishing what is unique about the human agent while respecting the deep commonalities across agents of varying complexity. The general principles of investment and stability govern agency at all levels of organization, and they find their ultimate grounding in the living system's need to sustain its own order in the face of entropy. The task of a naturalistic philosophy of action, then, is to trace the development of various forms of agency back to their metabolic roots. Such forms include (but are in no way limited to) skilled action, coordinated interpersonal activity, 'mental actions' such as calculating sums in one's head, and artifact use. All have received substantial philosophical attention, but they can now be addressed from a fresh perspective that respects the extent to which human action is a specialized form of a much broader phenomenon. Rational animals are still animals, after all.

Notes

1 H. Jonas, *The Phenomenon of Life: Toward a Philosophical Biology* (New York, NY: Harper & Row Publishers, Inc., 1966), p. 106.
2 A. Moreno and A. Lasa, 'From Basic Adaptivity to Early Mind', *Evolution and Cognition* 9:1 (2003), pp. 12–30.
3 Ibid.
4 X. Barandiaran, E. Di Paolo, and M. Rohde, 'Defining Agency: Individuality, Normativity, Asymmetry and Spatio-temporality in Action', *Adaptive Behavior* 17:5 (2009), pp. 367–386.

References

Adair, E.R. and Wright, B.A., 'Behavioral Thermoregulation in the Squirrel Monkey When Response Effort Is Varied', *The Journal of Comparative Physiology and Psychology* 90:2 (1976), pp. 179–184.

Anellos, A. and Moreno, A., 'Multicellular Agency: An Organizational View', *Biology and Philosophy* 30:3 (2015), pp. 333–357.

Anscombe, G.E.M., *Intention* (Oxford, England: Basil Blackwell, 1957).

Ashby, W.R., 'Principles of the Self-Organizing System', in H. Von Foerster and G.W. Zoph., Jr. (Eds.), *Principles of Self-organization: Transactions of the University of Illinois Symposium* (London, UK: Pergamon Press, 1962), pp. 255–278.

Barandiaran, X., Di Paolo, E., and Rohde, M., 'Defining Agency: Individuality, Normativity, Asymmetry and Spatio-temporality in Action', *Adaptive Behavior* 17:5 (2009), pp. 367–386.

Barandiaran, X.E. and Egbert, M.D., 'Norm-establishing and Norm-following in Autonomous Agency', *Artificial Life* 20:1 (2013), pp. 5–28.

Beer, R.D., 'Autopoiesis and Cognition in the Game of Life', *Artificial Life* 10:3 (2004), pp. 309–326.

——— 'The Cognitive Domain of a Glider in the Game of Life, *Artificial Life* 20:2 (2014), pp. 183–206.

Bickhard, M.H., 'The Dynamic Emergence of Representation', in H. Clapin, P. Staines and P. Slezak (Eds.), *Representation in Mind: New Approaches to Mental Representation* (Oxford, UK: Elsevier, 2004), pp. 79–90.

Bishop, J., *Natural Agency* (Cambridge, UK: Cambridge University Press, 1989).

Brand, M., 'Proximate Causation of Action', *Philosophical Perspectives* 3 (1970), pp. 423–442, on p. 425.

Bratman, M., *Intention, Plans and Practical Reason* (Stanford, CA: CSLI Publications, 1999).

Brooks, R.A., 'Intelligence without Representation', *Artificial Life* 47 (1991), pp. 139–159.

Brownlee, J., 'Autonomous Distributed Control in the Immune System Using Diffuse Feedback', *CIS Technical Report, 070329A* (March 2007), at http://citeseerx.ist.psu.edu/viewdoc/download?doi=10.1.1.71.1657&rep=rep1&type=pdf [accessed 15 May 2015].

Burge, T., 'Primitive Agency and Natural Norms', *Philosophy and Phenomenological Research* 79:2 (2009), pp. 251–278.

Chan, D.K., 'Non-intentional Actions', *American Philosophical Quarterly* 32:2 (1995), pp. 139–151.

Clark, A., 'Soft Selves and Ecological Control', in D. Spurrett, D. Ross, H. Kincaid and L. Stephens (Eds.), *Distributed Cognition and the Will* (Cambridge, MA: MIT Press, 2006), pp. 101–122.

Davidson, D., 'Actions, Reasons and Causes', *The Journal of Philosophy* 60:23 (1963), pp. 685–700.

——— *Essays on Actions and Events* (Oxford: Oxford University Press, 1980).

Dawkins, R., *The Selfish Gene* (Oxford: Oxford University Press, 1976).

Dennett, D., *Consciousness Explained* (New York, NY: Back Bay Books, 1991).

Dennett, D., 'Real Patterns', *The Journal of Philosophy* 99:1 (1991), pp. 27–51.

Di Nucci, E., 'Automatic Actions: Challenging Causalism', *Rationality Markets and Morals* 2:1 (2011), pp. 179–200.

Di Paolo, E., 'Autopoiesis, Adaptivity, Teleology, Agency', *Phenomenology and the Cognitive Sciences* 4 (2005), pp. 429–452.

Di Paolo, E., Rohde, M., and De Jaegher, H., 'Horizons for the Enactive Mind: Values, Social Interaction, and Play', in J. Stewart, O. Gapenne and E. Di Paolo (Eds.), *Enaction: Toward a New Paradigm in Cognitive Science* (Cambridge, MA: MIT Press, 2010), pp. 33–87.

Dretske, F., *Explaining Behavior* (Cambridge, MA: MIT Press, 1988).

Fodor, J., *The Language of Thought* (Cambridge: Harvard University Press, 1975).

Frankfurt, H., 'The Problem of Action', *American Philosophical Quarterly* 15 (1978), pp. 157–162.

Freeman, W.J., *Societies of Brains: A Study in the Neuroscience of Love and Hate* (Hillsdale, NJ: Erlbaum, 1995).

Froese, T. and Di Paolo, E., 'The Enactive Approach: Theoretical Sketches from Cell to Society', *Pragmatics in Cognition* 19 (2011), pp. 1–36.

Gersani, M., Abramsky, Z., and Falik, O., 'Density Dependent Habitat Selection in Plants', *Evolutionary Ecology* 12 (1998), pp. 223–234.

Gibson, J., *The Ecological Approach to Visual Perception* (Boston, MA: Houghton-Mifflin, 1979).

Glazier, J. and Libchaber, A., 'Quasi-periodicity and Dynamical Systems: An Experimentalist's View', *IEEE Transactions on Circuits and Systems* 35:7 (1998), pp. 790–809.

Haken, H., 'Synergetics: An Approach to Self-Organization', in E. Yates (Ed.), *Self-Organizing Systems: The Emergence of Order* (New York: Plenum Press, 1987), pp. 417–437.

Hochner, B., 'How Nervous Systems Evolve in Relation to their Embodiment: What We Can Learn from Octopuses and Other Molluscs', *Brain, Behavior and Evolution* 82 (2013), pp. 19–30.

Hooker, C., 'Introduction to the Philosophy of Complex Systems', in C. Hooker (Ed.), *Handbook of the Philosophy of Science, Volume 10: Philosophy of Complex Systems* (Waltham, MA: Elsevier B.V., 2011), pp. 3–90.

Hughes, S., 'Antelope Activate the Acacia's Alarm System', *New Scientist* 127 (1990), p. 19.

Hursthouse, R., 'Arational Actions', *The Journal of Philosophy* 88:2 (1991), pp. 57–68.

Ijspeert, A.J., 'A Connectionist Central Pattern Generator for the Aquatic and Terrestrial Gaits of a Simulated Salamander', *Biological Cybernetics* 84 (2000), pp. 331–348.

James, W., *The Principles of Psychology, Volumes 1 and 2* (Boston: Harvard University Press), note p. 489.

Jonas, H., *The Phenomenon of Life: Toward a Philosophical Biology* (New York, NY: Harper & Row Publishers, Inc., 1966).

Jones, D., 'What Do Animat Models Model?' *The Journal of Experimental and Theoretical Artificial Intelligence* 24 (2013), pp. 475–488.

Juarrero, A., *Dynamics in Action: Intentional Behavior as a Complex System* (Cambridge, MA: MIT Press, 1999).

Kant, I., *The Critique of Judgment* (Indianapolis, IN: Hackett Publishing Company, 1987).

Kelso, J.A., 'On the Oscillatory Basis of Movement', *Bulletin of the Psychonomic Society* 18 (1981), p. 63.

Kelso, J.A., *Dynamic Patterns: The Self-Organization of Brain and Behaviour* (Cambridge, MA: MIT Press, 1995).

Klarreich, E., 'Huygen's Clocks Revisited', *American Scientist* 90:4 (July–August 2002), at www.americanscientist.org/issues/pub/huygenss-clocks-revisited [accessed 15 March 2014].

Levy, G., Flash, T., and Hochner, B., 'Arm Coordination in Octopus Crawling Involves Unique Motor Control Strategies', *Current Biology* 25:9 (2015), pp. 1195–1200.

Luisi, P.L., 'Autopoiesis: A Review and a Reappraisal', *Naturwissenschaften* 90 (2003), pp. 49–59.

McConnaughay, K.D.M. and Bazzaz, F.A., 'Is Physical Space a Soil Resource?' *Ecology* 72 (1991), pp. 94–103.

Macnab, R.M. and Koshland, D.E., Jr., 'The Gradient-Sensing Mechanism in Bacterial Chemotaxis', *Proceedings of the Natural Academy of Sciences* 69:9 (1972), pp. 2509–2512.

Marchitti, C. and Della Sala, S., 'Disentangling the Alien and Anarchic Hand', *Cognitive Neuropsychiatry* 3:3 (1998), pp. 191–207.

Maturana, H.R. and Varela, F.J., 'Autopoiesis and Cognition: The Realization of the Living', in *Boston Studies in the Philosophy of Science, Volume 42* (Dordrecht: D. Reidel, 1980).

Mele, A., 'Intentional Action and Wayward Causal Chains: The Problem of Tertiary Deviance', *Philosophical Studies* 51:1 (1987), pp. 55–60.

——— 'Recent Work on Intentional Action', *American Philosophical Quarterly* 29 (1992), pp. 199–217.

Merleau-Ponty, M., *The Structure of Behavior* (Boston, MA: Beacon Press, 1963).

Mill, J.S., *A System of Logic* (London: John Murray Ltd., 1961).

Millikan, R., 'Biosemantics', *The Journal of Philosophy* 86 (1989), pp. 281–297.

——— *White Queen Psychology and Other Essays for Alice* (Cambridge, MA: MIT Press, 1993).

Moreno, A., Exteberria, A., and Umerez, J., 'The Autonomy of Biological Individuals and Artificial Models', *BioSystems* 91 (2008), pp. 309–319.

Moreno, A. and Lasa, A., 'From Basic Adaptivity to Early Mind', *Evolution and Cognition* 9:1 (2003), pp. 12–30.

Mrosovsky, N., *Rheostasis: The Physiology of Change* (Oxford, UK: Oxford University Press, 1990).

Neisser, U. (ed.), *The Perceived Self: Ecological and Interpersonal Sources of Self-Knowledge* (Cambridge, UK: Cambridge University Press, 1993).

Nicholson, D., 'The Return of the Organism as a Fundamental Explanatory Concept in Biology', *Philosophy Compass* 9:5 (2014), pp. 347–359.

Odling-Smee, J., *Niche Construction: The Neglected Process in Evolution* (Princeton: Princeton University Press, 2003).

O'Reagan, J.K. and Noe, A., 'A Sensorimotor Account of Vision and Visual Consciousness', *Behavioral and Brain Sciences* 24 (2001), pp. 939–1031, on pp. 940–941.

O'Shaughnessy, B., 'Trying (as the Mental 'Pineal Gland')', *The Journal of Philosophy* 70:3 (1973), pp. 365–386.

——— *The Will: A Dual Aspect Theory* (2nd ed., Cambridge, UK: Cambridge University Press, 2008).

Pacherie, E., 'The Content of Intentions', *Mind & Language* 15:4 (2000), pp. 400–432.

Penfield, W., *The Mystery of the Human Mind: A Critical Study of Consciousness and the Human Brain* (Princeton, NJ: Princeton University Press, 1978), p. 76.

Pettit, P., 'The Reality of Group Agents', in C. Mantzavinos (Ed.), *Philosophy of the Social Sciences: Philosophical Theory and Scientific Practice* (Cambridge, UK: Cambridge University Press, 2009), pp. 67–97.

Prigogine, I. and Stengers, I., *Order Out of Chaos* (New York, NY: Bantam, 1984).

Rapoport, A., 'Mathematical Aspects of General Systems Analysis', *General Systems* 11 (1966), pp. 3–11.

Rode, G., Charles, N., Perenin, M.T., Vighetto, A., Trillet, M., and Ainard, G., 'Partial Remission of Hemiplegia and Somatoparaphrenia through Vestibular Stimulation in a Case of Unilateral Neglect', *Cortex* 28:2 (1992), pp. 203–208, on p. 206.

Rossel, S., 'Binocular Spatial Localization in the Praying Mantis', *The Journal of Experimental Biology* 120 (1986), pp. 265–281, on pp. 274–275.

Ruiz-Mirazo, K. and Moreno, A., 'Basic Autonomy as a Fundamental Step in the Synthesis of Life', *Artificial Life* 10:3 (2004), pp. 235–259.

Russell, B., *The Analysis of Mind* (New York: Cosimo, Inc., 2004).

Sachs, T., Novoplansky, A., and Cohen, D., 'Plants as Competing Populations of Redundant Organs', *Plant, Cell and Environment* 16 (1993), pp. 765–770.

Sands, D., *Introduction to Crystallography* (Mineola, NY: Dover, 1994).

Schmidt, R.C., Carello, C., and Turvey, M.T., 'Phase Transitions and Critical Fluctuations in the Visual Coordination of Rhythmic Movement between People', *Journal of Experimental Psychology: Human Perception and Performance* 16 (1990), pp. 227–247.

Searle, J., *Intentionality: An Essay in the Philosophy of Mind* (New York: Cambridge University Press, 1983).

Thelen, E., 'Self-organization in Developmental Processes: Can Systems Approaches Work?' *Systems in Development: The Minnesota Symposia in Child Psychology* 22 (1989), pp. 555–591.

Thelen, E. and Fisher, D.M., 'The Organization of Spontaneous Leg Movements in Newborn Infants', *Journal of Motor Behavior* 15 (1983), pp. 353–377.

Thelen, E., Ridley-Johnson, R., and Fischer, D.M., 'Shifting Patterns of Bilateral Coordination and Lateral Dominance in the Leg Movements of Young Infants', *Developmental Psychology* 16 (1983), pp. 29–46.

Thelen, E. and Smith, L., *A Dynamic Systems Approach to the Development of Cognition and Action* (Cambridge, MA: MIT Press, 1994).

Thelen, E. and Ulrich, B.D., 'Hidden Skills: A Dynamic Systems Analysis of Treadmill Stepping During the First Year', *Monographs of the Society for Research in Child Development, Serial No. 223* 56:1 (1991), pp. 1–98.

Thompson, E., *Mind in Life: Biology, Phenomenology, and the Sciences of Mind* (Cambridge, MA: Belknap Press, 2007).

Tuller, B. and Kelso, J.A., 'Environmentally-specified Patterns of Movement Coordination in Normal and Split-brain Patients', *Experimental Brain Research* 75 (1989), pp. 306–316.

Varela, F., *Principles of Biological Autonomy* (Amsterdam: North-Holland, 1979).

Virgo, N., Egbert, M., and Froese, T., 'The Role of the Spatial Boundary in Autopoiesis', *Advances in Artificial Life: Darwin Meets Von Neumann* 5777 (2011), pp. 240–247.

Von Uexküll, J., 'A Stroll through the Worlds of Animals and Men', in K. Lashley (Ed.), *Instinctive Behavior* (New York: International Universities Press, 1934), pp. 5–80.

Weber, A. and Varela, F., 'Life after Kant: Natural Purposes and the Autopoietic Foundations of Biological Individuality', *Phenomenology and the Cognitive Sciences* 1:2 (2002), pp. 97–125.

Whelan, P., 'Control of Locomotion in the Decerebrate Cat', *Progress in Neurobiology* 49:5 (1996), pp. 481–515.

Wilson, G., 'Reasons as Causes for Action', in G. Holmstrom-Hintikka and R. Tuomela (Eds.), *Contemporary Action Theory* (Dordrecht, NL: Kluwer, 1997), pp. 65–82, on p. 146.

Winther, R., 'Part-Whole Science', *Synthese* 178 (2011), pp. 397–427.

Wittgenstein, L., *Philosophical Investigations* (Malden, MA: Blackwell Publishing, 1958).

Wright, L., 'Functions', *The Philosophical Review* 82:2 (1973), pp. 139–168.

Zullo, L., Sumbre, G., Agnisola, C., Flash, T., and Hochner, B., 'Nonsomatotopic Organization of the Higher Motor Centers in Octopus', *Current Biology* 19 (2009), pp. 1632–1636.

Index

action, as a natural kind 6–7; ascriptions to others 8, 36, 68; attributed to whole agents 7–8, 35–6, 94; causal theory of 2–3, 18–20; full-blooded 4, 5, 14, 15, 20, 26, 28, 31, 59, 98; intentional 1–4, 10, 14, 21, 26–8, 31, 59; 'mindless' 4, 27, 98, 99; non-intentional 4–6; phenomenology of 10–14, 17, 28, 95; primitive xi, 4–6, 7–15, 20–1, 24, 27, 29, 31, 34–5, 37–8, 43, 45–7, 50, 54, 83, 91, 94, 96, 98–100; properties 7–15; as spatiotemporal structures of process 89–91
adaptivity 83–5; asymmetry of 86; definition 84; homeostatic 85–6; supported by movement 87–8; not entailed by autopoiesis 83–4
affordances 11
alien hand syndrome (somatoparaphrenia) 10–11
amoeba xi, 35, 38, 39, 46
anarchic hand syndrome 12
Anscombe, G.E.M. 10
antirealism 36, 59, 62, 71
Ashby, W.R. 51, 59, 60
autocatalysis 69
autonomy 69–70, 80
autopoiesis 69–80; definition 70; function 72; insufficient for agency 83–4; role of boundary 70–1

Barandiaran, Xabier 77, 79, 83
Beer, Randall 74–6
behavior 42–7
Bernstein, Nicolai 56
Bickhard, Mark 44
Bishop, John 26
Brand, Myles 25

Bratman, Michael 3, 4
Brooks, Rodney 40
Burge, Tyler 5, 35–9, 42, 43, 46, 47

cat, decerebrate 40–1
causation, efficient 2, 47, 58, 91; top-down 61
centralized control 38–9
central nervous system: autonomy of 80, 98–9; role in coordinating action 39, 92
central pattern generator (cpg) 40
Clark, Andy 42
closure: operational 62, 68, 69, 87, 90, 91, 99; organizational 66, 69, 70, 71, 94
complex systems 51; dynamical analyses 51–3; human body as complex system 51, 53; self-organization 51–4
constraint: context-free 55–6; context-sensitive 50, 55, 91; entrainment 54–5, 57

Davidson, Donald 2–4, 19, 23
Dawkins, Richard 36
decerebrate cat 40–1
Dennett, Daniel 59, 60
deviant causal chains 18–20, 22, 25, 30; nanodeviance case 21–4, 27, 29, 30, 57, 58
Di Paolo, Ezequiel 84, 86
distributed control 40–2, 92
domain of interactions 74, 76, 77, 83, 88, 89, 94
Dretske, Fred 44–6
dynamical systems theory 51–3

Egbert, Matthew 77, 79, 83
embedding 59–60; and organism/environment "fit" 75
emergence 61, 69